QP
82.2
.P6
M4
V. 38

30-1

The Institute of Biology's
Studies in Biology no. 38

The Biology
of Pollution

by Kenneth Mellanby C.B.E., Sc.D., F.I.Biol.
Director, Monks Wood Experimental Station,
Honorary Professor of Biology, University of Leicester,
President of The Institute of Biology

Edward Arnold

© Kenneth Mellanby 1972

First published 1972
by Edward Arnold (Publishers) Limited,
25 Hill Street,
London, W1X 8LL

Reprinted 1974

Boards edition ISBN: 0 7131 2380 x
Paper edition ISBN: 0 7131 2381 8

Printed in Great Britain by
The Camelot Press Ltd, London and Southampton

General Preface to the Series

It is no longer possible for one textbook to cover the whole field of Biology and to remain sufficiently up to date. At the same time students at school, and indeed those in their first year at universities, must be contemporary in their biological outlook and know where the most important developments are taking place.

The Biological Education Committee, set up jointly by the Royal Society and the Institute of Biology, is sponsoring, therefore, the production of a series of booklets dealing with limited biological topics in which recent progress has been most rapid and important.

A feature of the series is that the booklets indicate as clearly as possible the methods that have been employed in elucidating the problems with which they deal. Wherever appropriate there are suggestions for practical work for the student. To ensure that each booklet is kept up to date, comments and questions about the contents may be sent to the author or the Institute.

1972

INSTITUTE OF BIOLOGY
41 Queen's Gate
London, S.W.7

Preface

Today we are all aware that pollution is a problem. We are bombarded with propaganda in books, in the press and on the radio and television. Much of this propaganda is of an extreme nature, proclaiming that we are poisoning ourselves and destroying our planet. Environmental damage is said to be caused by the population explosion and by our economic progress, and to be an inevitable byproduct of any increase of the Gross National Product. On the other hand politicians of all parties preach the gospel of growth, and tell us that without it any improvement in our environment is impossible. Which view are we to believe? There is a danger that the propagandist of doom will cry wolf too often, and so real dangers will be ignored. There is also a danger that the apostles of growth will so overwhelm the opposition that irreversible environmental damage will not be recognized. This booklet tries to show where real damage is caused by different types of pollution. I have written as a biologist, stressing the effects of pollution on living organisms, for I believe that this is the only way to recognize the real as opposed to the imaginary effects of toxic substances in our environment. On this knowledge only can a rational programme for pollution control be built.

Monks Wood, 1972   K.M.

Contents

What is Pollution?

1.1 Levels of pollution

Our world is full of poisonous substances. Many of these occur naturally, quite independently of any activity of man. Thus the vapours from an active volcano may contain so much sulphur that plants cannot grow nearby. Rivers flowing through forests may become deoxygenated because so much natural organic matter is deposited in them, and when this decomposes the results are similar to heavy contamination with raw, man-made, sewage. Mercury, occurring naturally in the ocean, may be concentrated by fish to levels which give concern to public health authorities. But when we consider pollution we usually refer to the presence of toxic materials introduced into our environment by man. This does not mean that only man-made pollution is harmful, though the suddenness of the changes induced by him are often more dramatic than the long term effects of naturally occurring poisons which we may have come to accept.

I consider that we should only speak of pollution where some effect, probably a harmful effect, can be recognized. Modern chemical methods are so sensitive that it is possible to detect traces of dangerous poisons everywhere. Our own bodies, even when we are in perfect health, contain quite substantial amounts of naturally occurring substances which are usually considered as poisons, including arsenic, mercury and other heavy metals. In addition we have picked up without suffering in health measurable amounts of man-made poisons like the insecticide DDT and industrial chemicals such as the polychlor biphenyls. Our bodies could be said to be 'polluted', at least by the man-made chemicals, but it is better to treat these traces as contaminants unless they can be shown to be having some characteristic effect.

However, the difference between harmful pollution and harmless contamination is not always a clear one. When a poison is present at a level where acute toxic effects can be recognized, this is clearly a case of gross pollution. We see this in a river receiving a massive amount of raw sewage, or near a brickworks pouring out sulphur and fluoride and killing the vegetation. It is more difficult to be sure that a lower level has no harmful effect. In the case of sewage or sulphur, which soon cease to be toxic, small amounts will usually disappear before any unrecognized damage can occur. However, some other poisons are cumulative. They may either be concentrated in the tissues of living animals or plants, or each minor exposure may have an additive effect. Thus a single exposure to lead, arsenic or DDT may do no recognizable harm, though a measurable amount of the poison may be stored in the body of the victim. Further exposures may produce an accumulation of the poison to a dangerous level. Chronic exposure to low levels of radiation may act rather differently, but with an apparently similar end result. Here each exposure, even the least intense, does *some* damage, even if this cannot be recognized. It is the summation of such effects which produces serious, and irreversible, results. In the case of the effects of cumulative chemical poisons which are concentrated in our tissues, permanent damage may not be inevitable and may be prevented if the toxic substances are eliminated from the body.

This difference between pollutants which are persistent and those which are not affects the whole subject of pollution control. The majority of pollutants are not persistent. In most cases all that need be done is to dilute the substance sufficiently so that it is below the level at which it is a recognized poison, and the problem is solved. The diluted poison will then be chemically changed into something harmless. Thus raw sewage can be run well into the open ocean, and it will break down harmlessly. Toxic sulphur dioxide gas may be blown up into the upper atmosphere and diluted to a harmless level, when it will soon become incorporated in substances such as ammonium sulphate which lack the original toxicity. It is of course possible for such plans to go wrong. Unusual currents following storms may deposit the sewage onto bathing beaches, and unusual weather may prevent the sulphur dioxide from being dispersed and diluted. But on the whole dilution and dispersal is a satisfactory means of controlling the majority of non-persistent pollutants.

The persistent pollutants, often spoken of today with more accuracy than elegance as 'non-biodegradable', pose a very different problem. When they are diluted to harmless levels, they may still remain within our environment, with the possibility that they may be concentrated, perhaps by living organisms, until they are once again present at dangerous levels. For instance, a fish may concentrate some of the organochlorine insecticides from the water in which it swims by a factor of as much as 10 000. But there is a limit to this process, and eventually dilution will always produce conditions where insecticides, for example, are lost from the body and not increased in amount. Also few of even the most persistent chemicals remain unchanged indefinitely. So the fears of permanent global pollution by man-made chemicals are often overestimated. Nevertheless these persistent pollutants do present very difficult problems to those who have to deal with them, and future industrial development is likely to produce more rather than fewer substances which fall into this category.

1.2   Population and pollution

Pollution becomes a more serious problem as our population increases, and as our industrialization becomes more intense. Primitive man, living in small numbers, had little adverse effect on his environment. His sewage could be harmlessly absorbed by the rivers, his smoke soon disappeared into the atmosphere. It was when the population grew and when he came to live in cities that his wastes began to make their impact by poisoning the waters and the air. Then industrial development took place, causing serious damage as poisonous substances were directed by man into the wrong situations. It should be remembered that in many cases man did not create the poisonous chemicals. We have large areas where the soil is made sterile by the presence of high levels of lead, zinc, copper or arsenic. These substances all occurred before they were mined and transported to the industrial plants. In their original location, buried beneath the surface, they usually did little harm, though cases of natural poisoning of the soil and the vegetation are not uncommon, but dispersed by industry the effects were greatly increased. Industry also synthesizes new pollutants, such as the organochlorine insecticides and the polychlor biphenyls which have already been mentioned.

Clearly a great fear must be that we shall be unable to contain our pollution. The population of the world is increasing, and is expected to double by the year 2000. So far industrial development is mainly restricted to a few developed countries, but the whole world hopes (probably vainly) to raise its standards of life to those of Western Europe and North America. More people and more industry will pose great problems of food supplies, of power and of waste disposal. The suggestion that world population will double every twenty or thirty years until there is 'standing room only' is clearly nonsense; long before that happens our population, like that of any other animal, will be controlled by some other means. This means could be pollution. It could be pestilence. It could be a nuclear holocaust. It is clear that if civilization is to survive, the growth of the human population of the world must, in some way, be controlled. This, in the long run, is man's greatest problem. In the meantime life can only be made tolerable if pollution is controlled, and if the irreversible degradation of our environment is prevented. But the effects of even the most efficient pollution control will eventually be nullified if the world's population continues to increase at the present rate. Fortunately for me population control, though of such paramount importance, is not the subject of this booklet.

1.3 Biological effects of pollution

I have insisted that the term 'pollution' should only be used when there is actual or potential damage to man or to the environment. Physical damage from air pollution to buildings and to metals can be easily demonstrated, and corrosion by water pollution can also be shown. However, I believe that it is the biological effects of pollution that are of paramount importance. These can often be detected before any physical or chemical effects are easily visible. In many cases it is the danger to human health, an obviously biological parameter, which is the accepted reason for pollution control.

At the beginning of this chapter I tried to distinguish between real pollution and the harmless presence of potentially toxic substances at such low levels that no harm was done, and I showed that this distinction was sometimes difficult to establish. Here the biologist has a contribution to make. I believe that we should try to prevent any pollutant reaching a level where any biological reaction can be demonstrated (e.g. an apparently innocuous but nevertheless definite change in cellular metabolism) even if this reaction has not been shown to be harmful. So often in the past the chronic effects of toxic substances, as of X-rays, have been underestimated. This does widen my definition of what I would call pollution, to include the levels of potentially dangerous substances which can be shown to have some recognizable biological effect.

This brings us to the question of standards. I have already mentioned that fish can concentrate some poisonous substances, such as organo-chlorine insecticides, by a factor of 10 000. Thus water containing DDT at a level of one part in 100 000 000 would be lethal to some fish, whereas it could be drunk with impunity by man, who could only obtain a hundredth of a milligram for every litre imbibed. To the water authority this water could be treated as 'pure', whereas to the angler it would be seen to be grossly polluted. If we are to be concerned with our whole environment, then we must adopt the higher of these two standards. To implement these at present may not always be practicable, but in the long run the full restoration of all damage should be our goal. The point here is that such damage must be easily demonstrated. At the same time it is important not to waste our energies on trying to get rid of contaminants where absolutely no biological effects on man, on plants or on animals can be demonstrated.

My purpose in this booklet is, then, to concentrate on the biological effects of environmental pollution. I shall not deal with some of the hypothetical changes which the more alarmist writers foretell as imminent dangers likely to destroy the whole climatic balance of the world, or as probable causes of the destruction of our atmosphere. Thus it is suggested that the increase in the levels of carbon dioxide in the atmosphere, resulting from the increasingly rapid consumption of fossil fuels, will trap an increasing amount of the heat re-radiated from the earth's surface. This will produce a so-called 'greenhouse-effect' and raise the world temperature, possibly by several degrees. This will melt the polar ice, and that which is not floating will contribute to a rise in sea level which, in some calculations, could flood many of our cities and the surrounding lands up to, perhaps, a hundred metres above the present high tide mark. On the other hand the increasing production of smoke from industries, particularly in developing countries, could cut down the radiation reaching the earth from the sun and precipitate a new ice age with a consequent fall in sea level. Supersonic planes, flying at high levels, producing smoke and water vapour in the thin atmosphere, and using up great quantities of oxygen, have been cited as possible causes of climatic modification. Any of these changes would have devastating effects with their biological components. Clearly if there is any chance of their occurrence we must try to be prepared, and in the meantime must monitor world changes in climate to see what is really happening. I am personally not acutely worried on this account, as the most competent authorities seem to consider that the effects of man's efforts on the world's climate are likely to be much less than the fluctuations which occur without his intervention. It should be remembered that we had an ice age a few thousand years ago, before man had made any contribution to global pollution.

The possible effects of pollution on the composition of the atmosphere are also widely discussed. We know that the air once contained little oxygen, and some writers suggest that a return to this state, and our suffocation as a result, is imminent. Oxygen is used up by the respiration of animals and plants, and by industrial processes. The supply is maintained by the photosynthesis of plants on land and in the sea. We are destroying our forests, and may be poisoning the oceans and therefore the phyto-plankton which produces so much oxygen. Here again I do not think we have reason for panic. The evidence that the phyto-plankton is in danger from pesticides is not convincing (see Chapter 7). Even if no oxygen were being replaced (and most if not all is, at present, being put back) supplies in the atmosphere are sufficient for several hundred years. No doubt, uncontrolled growth of the world population and of industry would, eventually, affect the climate and the atmosphere, but long before this became effective other catastrophes would have taken their toll.

Air Pollution

2.1 Atmospheric constituents

The atmosphere of the troposphere—the layer of air supporting all plant and animal life and covering the globe to a height of approximately 10 kilometers—consists of 21 per cent oxygen, 78 per cent nitrogen, and 0.033 per cent carbon dioxide (percentages in dry air). It also includes about one per cent of argon, neon, helium and other inert gases which, as far as most living organisms are concerned, act much like nitrogen. Water vapour is present in amounts varying from a fraction of one per cent in cold dry air to three or four per cent during the wet season in the tropics. The moisture of the air is clearly of considerable importance to animals and plants, and its variations have biological effects. In so far as man affects the atmospheric humidity this factor could, if unfavourable, be considered to be a form of pollution. However, except in very local situations (e.g. over-dry air in some centrally-heated buildings, condensation of steam from super-saturated exhaust gases), man's influence is negligible, and for our purposes can be ignored.

Pollution seldom affects the basic constitution of the air, for instance by reducing the oxygen available for respiration. A significant reduction can occur in a confined space like a mine, but not in a room like a crowded lecture theatre with the windows closed. Altitude has a far greater effect. On the top of Ben Nevis the oxygen level is reduced by about a sixth, a fall that is hardly detected by a healthy individual. At greater heights, as in unpressurized aeroplanes, oxygen deficiency can be more serious, and extra oxygen was needed to climb Mount Everest.

Most pollutants are added, usually in quite small amounts, to normal air. The intensity of pollution can be expressed in various ways. We often speak of the weight of the pollutant in a given volume of air, thus SO_2 may occur as one milligram (or 1000 μg) per cubic metre. This method of expression can be used for both gaseous and solid pollutants. However, for gases it is common also to speak of parts per million, which means the number of cubic centimetres of the gas in a cubic metre. This may be confusing, as for other pollutants in, for instance, body tissues, one part per million means one milligram per kilogram of body weight. To transform parts per million in the air to mg per cubic metre, the following formula may be used.

$$\text{ppm} = \text{mg m}^{-3} \times \frac{22.4}{\text{molecular weight of pollutant gas}}$$

Thus normal air has 0.033 per cent CO_2 (molecular weight 44). This can also be expressed as 330 ppm or 660 mg m^{-3}. Similarly if SO_2 (molecular weight 64) occurs at a level of 0.3 ppm, this can also be expressed as 1 mg m^{-3}.

The effects of most pollutants depend on their concentration. If this is sufficiently high, acute effects will be seen. As far as man is concerned, levels of emission into the air he breathes are usually well below those where acute effects are produced; places with higher levels of pollution are purposely avoided. Chronic damage may occur, but this may be difficult to recognize, and may only be detected when epidemiological data from large populations are carefully analyzed. Some plants are much more susceptible than man, and may show obvious symptoms of phyto-toxicity when animals appear, at least for a time, to be unharmed. The use

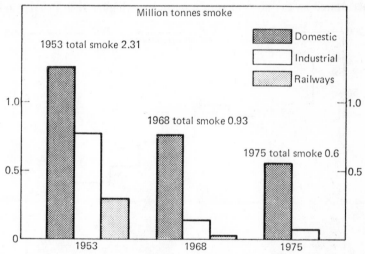

Fig. 2–1 Smoke emissions in the UK for 1953 and 1968, with a prediction for 1975. (From ROYAL COMMISSION ON ENVIRONMENTAL POLLUTION, 1971)

of such susceptible plants as indicators of pollution may be useful, parti-cularly as they may integrate the results of exposures at different levels over a period of time. (See p. 12.)

2.2 Fossil fuels

The burning of fossil fuels, particularly coal and oil, is the greatest cause of air pollution. Coal produces enormous numbers of particles of different sizes. The larger ones make up the dust, the smaller, smoke. The dust is usually deposited near to its source, and until recently as much as a kilogram fell in a year on each square metre near to some factories. The dust made our cities filthy, it helped to make them dark, its very weight harmed the vegetation and choked people breathing it in. However, the larger particles were probably excluded from the lungs by the very process which made breathing uncomfortable, and the harmful effects were therefore less than might have been expected. Fortunately in Britain dust production by industry is now only a tiny fraction of that emitted twenty years ago (Fig. 2–1). The problem still continues in other parts of the world, but it is clearly one which we could solve. In any case the dust does not remain long in the air, and it is usually deposited over a restricted area.

The smaller smoke particles remain longer in the air. They are breathed into the lungs, where many remain permanently to blacken the tissues. It is usually assumed that exposure to smoky air is harmful to man, but there is very little evidence for this view except, as will be shown later, when the smoke is a constituent of fog. Of course in fog the pollution is trapped and concentrated so the level of smoke particles is higher than occurs under

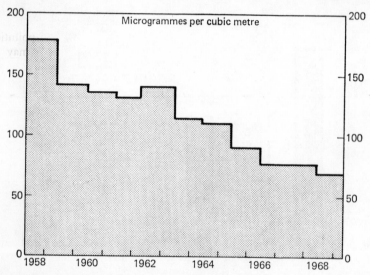

Fig. 2–2 Average smoke concentration near ground level in the UK, 1958–1968. (From ROYAL COMMISSION ON ENVIRONMENTAL POLLUTION, 1971)

other weather conditions (Figs. 2–2 and 2–3). There are few experimental or epidemiological results which quantify the harmful effects of smoke even at high concentrations, except in the case of that produced by burning tobacco, where the levels of pollution are clearly much greater than would be tolerated from any other type of combustion.

2.3 Sulphur pollution

The other main pollutant from fossil fuels is sulphur, mainly emitted as the gas sulphur dioxide, of which some 6 000 000 tonnes are produced in

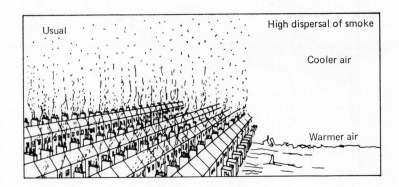

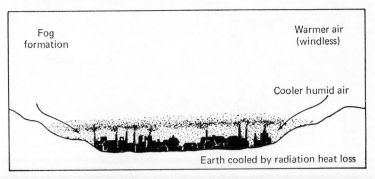

Fig. 2–3 The mechanism of temperature inversion. These illustrations show how change in air temperature can trap pollutants in a dense fog close to the ground. Many industrial towns in Britain are by their situation especially exposed to this hazard. (From COMMITTEE OF THE ROYAL COLLEGE OF PHYSICIANS, 1970)

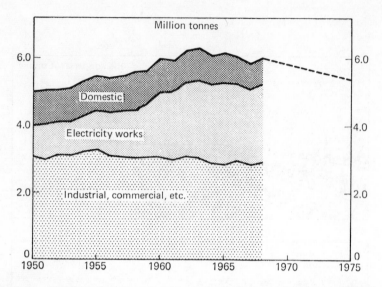

Fig. 2-4   Sulphur dioxide emissions in the UK, 1950–1968 (with prediction to 1975, based on *Fuel Policy*—Cmnd. 3438, 1967). (From ROYAL COMMISSION ON ENVIRONMENTAL POLLUTION, 1971)

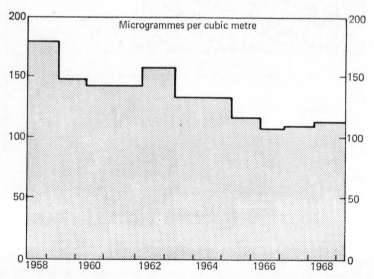

Fig. 2-5   Average sulphur dioxide concentrations near ground level in the UK, 1958–1968. (From ROYAL COMMISSION ON ENVIRONMENTAL POLLUTION, 1971)

Britain annually (Figs. 2–4, 2–5, 2–6). High levels in industrial countries are man-made, but global production of sulphur dioxide is nearly seventy per cent from natural sources (e.g. decomposition of vegetation in swamps). In addition there is no evidence of a global problem of sulphur production as background levels are well below 0.01 ppm and do not approach the threshold where effects on animals or plants begin to appear. An important point is that sulphur dioxide as such does not remain in the atmosphere for

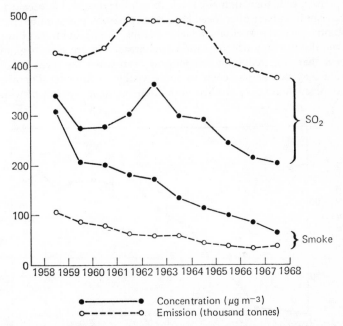

Fig. 2–6 Changes in the emission of smoke and sulphur dioxide and their concentration in London air. (From COMMITTEE OF THE ROYAL COLLEGE OF PHYSICIANS, 1970) In London, implementation of the Clean Air Act of 1956 has markedly reduced the amount of smoke emitted and lowered its concentration. The emission of sulphur dioxide in the air has changed little, but better dispersion has also produced lower concentration of this pollutant in London air.

very long, as it usually combines with the excess of ammonia present to produce ammonium sulphate, a substance which does not have its toxic properties. This transformation may occur in minutes or hours under warm, moist conditions, and is always likely to be virtually complete within a couple of days. Thus though sulphur pollution may cross international frontiers, reports of dangerous levels being transported over the breadth of a continent are probably exaggerated.

We know more about the toxicity of sulphur dioxide than we do of that of smoke. Experiments with human volunteers show no recognisable effects with levels below 1 ppm (3 mg m^{-3}), a concentration seldom found in any British city. Higher levels sometimes occur in industrial plants, where the workers may assert that they relieve the symptoms of nasal congestion accompanying a cold. This apparently beneficial effect should, however, be treated with caution, for it suggests a definite physiological effect which, if prolonged, could easily cause serious damage.

There is more evidence of damage to plants, which appear to be much more susceptible than animals: levels of sulphur dioxide between 0.1 and 1 part per million have often been shown to cause obvious symptoms such as leaf blotching and reductions in yield of crops. Coniferous trees grow poorly or even die in many urban or industrial areas. We are now beginning to suspect that crop reductions may occur even when the levels of sulphur dioxide may not be sufficient to cause visible symptoms. The group of plants most thoroughly studied is the lichens, many species of which are

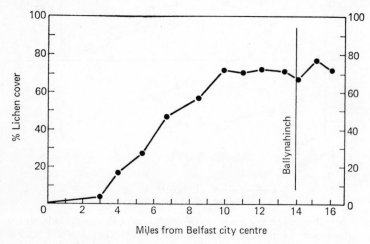

Fig. 2-7   Increase of lichen cover outside the city of Belfast. (After A. F. Fenton, from MELLANBY, 1967)

absent from urban areas because of their extreme susceptibility to sulphur dioxide damage; they clearly serve as very sensitive indicators. Also, if they are completely prevented from colonizing an area where sulphur dioxide levels are as low as 0.05 ppm, this concentration of the gas is clearly having a biological effect, and it would therefore be rash to state dogmatically that other forms of life, including man, could not be somewhat affected (Fig. 2-7).

There is, however, one other instance where sulphur pollution has been welcomed. Some of our soils are deficient in this element, and better yields, particularly of grass, have been attributed to its deposition from the atmo-

sphere. Here again I would treat this situation with caution. A physiological effect, even a beneficial one, occurring in one area, could mean that a
very small increase in sulphur level would overstep the limit and cause at
least chronic damage nearby. A pollutant which has *no* effect is more safely
tolerated than one which may, apparently, be beneficial under particularly
favourable circumstances. As existing sulphur dioxide levels in and near
our cities certainly harm susceptible plants, even if man is not apparently
at risk, I consider that we should make greater efforts to control its emission
even though dispersal by means of high chimneys has resulted in a lower
ground level concentration in most urban areas of Britain in recent years,
notwithstanding a continuing high production of the gas. Recent reports
from the United States that sulphur may be recovered, economically, by
extraction from flue gases, are encouraging. Such processes were being
actively developed in Britain in the 1960s, but work was discontinued
when world sulphur prices fell greatly below those where the existing
cost of extraction could compete. In the long run I believe that we will be
able to reduce sulphur pollution in this way at a reasonable cost, if not at an
actual profit.

Taken separately, it seems that the main pollutants from burning coal
can harm man and his environment, but that this is not often very serious.
The main damage, particularly to man, occurs when smoke contributes to
the production of fog or 'smog', as happened in London in December
1952. Figure 2–8 shows how the levels of smoke and sulphur dioxide
increased during the period 5 to 9 December, and how the deaths more than
trebled. This was the result of the weather, when the air was still and there
was a temperature inversion which trapped the pollutants and concentrated them in the lower levels of the atmosphere. The total number of
deaths, mainly from bronchitis, pneumonia and other diseases of lungs,
and from heart diseases, was, taking the period of the smog and its after-
math, something more than 4000 in excess of the mortality which would
have otherwise been expected during the same period.

We do not know exactly what caused these deaths. The levels of sulphur
dioxide, with a maximum of 0.75 ppm, and of smoke, which reached
1.5 mg m^{-3}, would not themselves seem sufficient to have such serious
effects. Other contaminants were present, e.g. sulphuric acid and oxides of
nitrogen, and these are known to be toxic but hardly at such low levels.
Man is known to suffer most when it is cold, as it was at this time. The
mortality is generally explained as the result of a combination of these
various factors.

The death of some 4000 individuals was very serious, but it only corres-
ponded to less than one per 2000 of the population at risk, which makes its
investigation difficult. Thus in a controlled animal experiment with a
similar mortality, the numbers used would have to be in the region of
100 000 in both the experimental and control batches if a significant result
were to be probable. It is therefore not surprising that we are still uncertain
of the precise effects of the various pollutants.

2.4   The Clean Air Act

The smog of 1952 stimulated governmental action which resulted, in 1956, in the passing of the Clean Air Act. Since then it has been possible to control the output of pollutants, particularly of smoke. In fact as industry was changing over from coal, with its high labour costs, to more

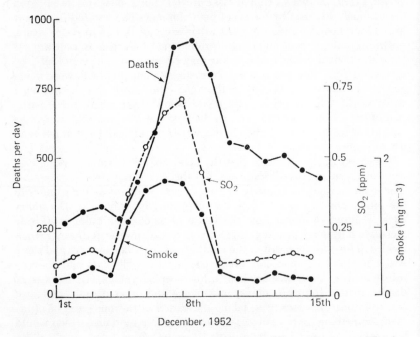

Fig. 2–8   Death and pollution levels in the fog of December 1952. The fog lasted from 5 December to 9 December. The graphs show the increased number of deaths during this period and also the rise in the amounts of sulphur dioxide (SO_2) and smoke in the air. The figures for death were provided by the Registrar General's Office. The pollution levels were measured at twelve different stations in London. (From COMMITTEE OF THE ROYAL COLLEGE OF PHYSICIANS, 1970)

convenient fuels like oil which do not produce much smoke, some improvement would have occurred anyway, but the legislation has caused more conservative firms to follow suit. This act also make it possible for local authorities to set up 'smokeless zones' where the domestic use of raw coal is prohibited. The improvement in the conditions of London and some other cities has been remarkable, though in the industrial North of England it has been disappointingly slow. In recent years when the weather pattern has resembled that of December 1952 there has been little or no smog, and no substantial increases in the death rate have been observed (Fig. 2–9).

Petrol and diesel engines are blamed for much air pollution. Lorries are often seen to belch black smoke from their exhausts, and motor cars are known to pass out toxic substances which include carbon monoxide, oxides of nitrogen, ozone, unburned hydrocarbons and lead. The difficult question is to quantify the danger to man and to the environment of these pollutants.

A properly maintained diesel engine produces a negligible amount of pollution. The black smoke so commonly seen is the result of poor maintenance or of deliberate abuse of the engine. Legislation exists which

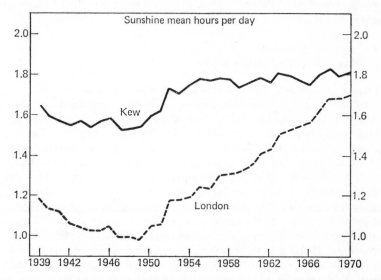

Fig. 2–9 Trend of winter sunshine (December to February) at London Weather Centre and Kew Observatory. (From ROYAL COMMISSION ON ENVIRONMENTAL POLLUTION, 1971)

could put an end to diesel smoke, but it is obviously not enforced. However, even at present levels diesel fumes do little damage to the health of man, to animals and even plants, even if they are so very unaesthetic.

At high levels, the pollutants from petrol engines are obviously dangerous. Death from carbon monoxide poisoning commonly occurs when a car is kept running in a closed garage. The level in busy city streets has given rise to much concern, because carbon monoxide has such a high affinity for haemoglobin in blood, producing carboxyhaemoglobin which interferes with the transport of oxygen. However, surveys in heavy traffic have seldom shown levels of carboxyhaemoglobin of more than 4 per cent, and this has not been conclusively shown to have any adverse effect on the behaviour or the efficiency of the victim. A normal individual is unlikely to be affected, though someone suffering from anaemia might show appreciable symptoms of oxygen deficiency. Much higher levels of carboxyhaemoglobin are commonly found in the blood of cigarette smokers without obvious effects. Concern is expressed lest carbon monoxide levels in cities should rise if the doubling of motor vehicles foretold for Britain in the next 15 years occurs. As most of our city streets are already saturated with stationary cars in traffic jams, the air can hardly be polluted more in these most polluted areas. The bulk of the additional cars will have to find space in other places. Nevertheless I do not think that we should accept the present levels of carbon monoxide, even if it cannot be proved that they are doing a great deal of damage. Some priority should therefore be given to engine improvements which cut down emission of this substance.

2.5   Photochemical smog

Oxides of nitrogen, which are produced as nitrogen is 'burned' in the car engine, are also poisonous in high concentrations, but these are seldom reached in air polluted by exhaust gases. In industrial accidents nitrogen dioxide may cause death, with severe lung damage in survivors. Levels of 5 ppm are considered safe in industry; the highest level reported in London is about 0.2 ppm, and usual levels are only a fraction of this. The margin of safety would therefore seem adequate. The concern over these substances arises from the production of what is called 'photochemical smog'. It should be stressed that this has no resemblance, except that visibility is somewhat obscured, to the smog resulting from smoke described earlier in this chapter, and it is perhaps unfortunate that the same word is used. Photochemical smog is most notorious in California. It results from the presence of many automobiles producing the exhaust gases already mentioned. When there is a temperature inversion this traps the pollutants at a high concentration in the air near the ground. Even when this occurs little harmful effect can be demonstrated unless there is a high intensity of sunshine. Under these conditions peroxacyl nitrates ('PAN') are produced. They are substances with a remarkable degree of biological activity, in that concentrations of a much lower magnitude than we have so far been considering, in the region of parts per hundred million (i.e. $10 \ \mu dm^3$ m^{-3}) cause visible damage to agricultural crops, and even lower concentrations ($1 \ \mu dm^3 \ m^{-3}$) cause severe eye irritation in man. Chronic exposure to concentrations even below these may also have harmful effects.

Photochemical smog is a serious problem in many parts of the world. California is the region most often quoted as being affected, but it can occur anywhere where there are many petrol engines, temperature inversions and brilliant sunlight. It has been reported in other parts of America, from Japan, and, occasionally, from Europe. It is said to occur in the Netherlands, though so far never at a Californian intensity. Although there are reports suggesting some production of PAN in Britain, there have been no instances which have merited the description of smog, and most authorities think that it is unlikely ever to be produced at dangerous levels.

There is one other reason why photochemical smog is unlikely in Britain. It has been shown that sulphur dioxide actually prevents its production; this is said to be the reason for the comparative rarity of smog in New York. The places in Britain where smog might develop are also those with most SO_2. However, this is hardly a good reason for delaying controls of sulphur pollution.

In the United States, efforts are being made to prevent photochemical smog from developing. Car exhausts are to be fitted with appliances to cut down the emission of the dangerous pollutants. It is likely that we will follow suit, even if smog is unlikely to be serious in Britain. If we can afford such precautions, a little extra—if somewhat unnecessary—purity will do no harm.

2.6 Lead pollution

The final air pollutant emitted by cars is lead. Tetraethyl lead is added to petrol for reasons of efficiency. One disadvantage of this is that the lead makes control of other pollutants by catalytical methods unpracticable, as the lead 'kills' the catalyst. It is also a pollutant in its own right. Vegetation near busy roads may contain as much as 500 ppm (by weight) of lead, and as such it is unsuitable for animal or human food. The pollution seldom stretches far from the road, and pollution of crops has seldom been reported. The most serious cause for concern is the lead which may be given out in aerosol form and is then breathed by city dwellers. So far surveys have been moderately encouraging. Men who worked in garages in America and who had the maximum exposure had levels of lead in their blood of 60 μg 100cm^{-3} in only a handful of cases, and this is below the danger level set by the health authorities. Other city dwellers, including traffic policemen, had even smaller amounts. However, I think that our standards may not be sufficiently rigorous, and I would support moves to remove lead as an additive to petrol, provided that it is not replaced by something more dangerous. There is a possibility that lead-free petrol would contain more benzene, which could be dangerous to those handling the liquid.

Concern has been expressed in America that the nutrient salts produced when the nitrates resulting from the natural detoxication of the nitrogen oxides produced by petrol engines, and the ammonium sulphate produced from the sulphur dioxide released from flue gases may give rise to eutrophication (see p. 27), and thus contribute to land and water pollution. The amounts of these nutrients are said to be equivalent to a substantial fraction, perhaps a third, of the amount of similar chemicals used in agriculture. It must be remembered, however, that these fertilizers are used on only a small fraction of the total area of the United States. I think that, at present at least, these 'pollutant' nutrients spread over the whole country will have little effect, as at most only one or two kilograms will fall on each hectare, and this is below the level when any effect can be detected.

2.7   The Alkali Act and Alkali Inspectorate

Today smoke, sulphur dioxide and the pollution from vehicles cause most concern to the public. Formerly most damage appeared to be caused by emissions from industrial sources. The first Alkali Act was passed in 1867 primarily to control the emission of hydrochloric acid from alkali works. There are still many areas around the sites of old factories and smelters where the ground is made sterile by the accumulation of old pollution by copper, zinc and arsenic. There has been an undoubted improvement in Britain, and for this the Alkali Inspectorate deserves much of the credit. This independent though government organized body has the unusual duty of controlling specified pollution by 'the best practicable means'. This system is often criticized by those who think we should have rigid, legal minimum levels to be permitted for each pollutant. There is little doubt that had this course been adopted from the outset, Britain would today be much dirtier than it is, for standards, to be universally applicable, would have been set for the worst areas and would have been lower than those enforceable in the best. Also there would have been a tendency to try to keep emissions just below the prohibited level. There have been times when the Inspectorate have been criticized for not being tough enough and for permitting old factories to continue to pollute the air, but the total end result has been a lesson to other countries with more rigid but less easily enforceable schemes. Old factories go out of use, and are then replaced by those with higher standards, designed to fulfil the more stringent requirements of the Alkali Inspectorate. This ensures a steady improvement of environmental quality.

Within the walls of some factories, higher levels of pollution sometimes continue. Here levels (as for nitrogen dioxide) may be monitored to prevent damage to workers. Masks may be worn to protect them from such substances as asbestos, the dust of which could otherwise produce serious bronchial illness and the possibility of lung cancer.

Some asbestos dust is carried by the air in cities, probably as the result of the wear of car brakes. It has been suggested that this dust presents a health hazard, though existing evidence suggests that it never reaches the threshold of the danger level.

The most serious industrial pollution today comes from brickworks and aluminium smelters. Sulphur dioxide from older brickworks with insufficiently high chimneys may damage crops in the vicinity. Newer works dispose of the gas higher and more efficiently.

Fluoride even at levels of 0.1 ppm or lower can cause serious damage to plants. Thus in Norway aluminium smelters at the bottom of deep valleys may be surrounded by many square kilometres of dead and dying coniferous trees. In Britain there have been reports of fluoride damage to agricultural crops, but the most serious effects have been on cattle. When fluoride is deposited on pasture, the grass concentrates the pollutant, and when it is eaten the animals are affected. Mild poisoning mottles the teeth, while in severe cases the skeletal bones are softened and eventually the animals die. Most modern smelters have very high chimneys so this local damage is prevented. But near brickworks fluoride poisoning can still be detected, and this danger prevents cattle being kept in some areas.

2.8   Smoking as pollution

It is clear that most other types of air pollution pale into insignificance before the one really serious form—cigarette smoke. There is now no doubt that much lung cancer and various bronchial and cardiac illnesses are associated with smoking. As has already been mentioned, higher levels of carbon monoxide are found in the blood of a smoker than in those exposed to the highest intensities of traffic fumes. The amount of particulate matter tolerated (often unwillingly) by the non-smoker surrounded by smokers is often far greater than that found near the smokiest coal fire—and these are illegal in many parts of the country. The particulate matter actually inhaled by smokers is of a different order of magnitude from that from most other objectionable combustions. Although the relationship between smoking and many forms of illness is clearly established, the actual mechanism causing this morbidity is still not fully understood. Levels of cancer differ in rural and urban areas, in different countries and between individuals of different nationalities. Nevertheless the general picture is quite clear. I am often surprised to find that some of those who attack air pollution, perhaps from motor cars, with such vigour, are still prepared to endanger their own health and to irritate their associates by continuing with this much more dangerous practice.

Water Pollution 3

Many of the inland waters of Britain suffer from gross pollution, by sewage, industrial effluents and the products of modern agriculture. Some 2000 kilometres of river are described as 'grossly polluted' so that the water cannot be used, even after purification, for domestic supplies or even, in some cases, by industry. This figure of 2000 km may seem not very serious, when compared, as it sometimes is, with the total length of our river system which is given as 32 000 km. In fact the polluted stretches include most of the lower portions of the larger rivers, and of the parts of those rivers on whose banks a substantial part of the total population of Britain lives. This means that a much greater proportion of the flowing water is polluted than figures would, at first glance, indicate.

3.1 Oxygen levels

However, many of the rivers which are considered to be 'unpolluted' are in fact suffering from the results of man's activities, and they are, from the ecological point of view, degraded. I have already said that we should only use the term pollution when some harmful effect, often a biological effect, can be shown. A change in the flora and fauna is such an effect, and if the change is for the worse, then this is surely justly described as a symptom of pollution. Even a moderately polluted river can be rendered safe for human use; such rivers are sometimes classified as 'clean' by the water industry.

I have shown that air, no matter how heavily polluted, still retains the normal proportions of its major constituents, nitrogen and oxygen, and so will almost always contain enough oxygen to support life. Polluted water may be totally deoxygenated, so that most aquatic forms of life may be unable to exist in it. At best water only contains, in solution, rather a low level of oxygen. Thus at 20 °C a litre of saturated water only contains some 6 cm^3 of oxygen while a litre of air contains over 200 cm^3. Even an unpolluted stream or pond is probably only saturated during the day when plants are engaged in photosynthesis and liberating oxygen; at night, when the plants, like the animals, are using it up, the oxygen levels may fall substantially.

The warmer the water, the less oxygen it is able to hold in solution. Thus lack of oxygen is one of the factors in thermal pollution (see Chapter 4). The difficulty for fish and invertebrate life is that the higher the temperature within the range where life is possible the greater is the need for oxygen, and the lower is the supply. This is a reason why a change in the temperature of a river or lake may alter the whole balance and range of species present. It should also be noted that water containing toxic pollutants usually has a low oxygen level. This means that fish must breathe very large volumes of water to obtain their oxygen. The water is taken in through the mouth and passed out over the gills. These absorb the oxygen; they may also absorb the poisonous pollutants. So the lower the level of dissolved oxygen, the higher the dose of pollutant likely to be absorbed.

The commonest type of water pollution is by organic matter such as sewage. This has the effect of stimulating bacterial and fungal growth, and

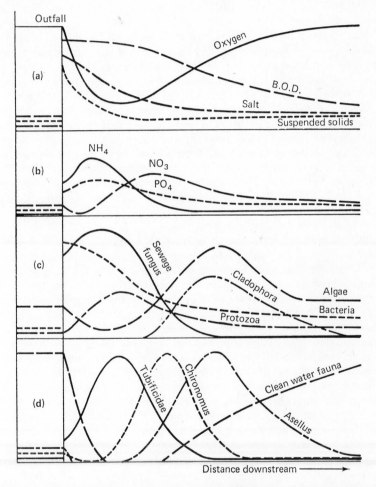

Fig. 3-1 Diagrammatic presentation of the effects of an organic effluent on a river and the changes as one passes downstream from the outfall. **(a)** and **(b)** physical and chemical changes. Most recently installed plants tend to reduce NH_4 and increase NO_3 levels, **(c)** Changes in micro-organisms, **(d)** Changes in larger animals. (From MELLANBY, 1967)

these processes absorb oxygen and so de-oxygenate the water. The organic matter is usually measured by its so-called Biochemical Oxygen Demand (B.O.D.). This is based on a test where the ability of polluted water to absorb oxygen is measured—the greater the oxygen demand, the more polluted the sample. The relationship between B.O.D. and the oxygen level in a river, just below an outfall where sewage or other organic matter enters it, is shown in Fig. 3-1. This example shows the chemical, physical and biological effects of a moderately severe pollution, but one in which the self-purifying processes are sufficient to overcome the pollution so that the river is eventually restored to the conditions which existed above the outfall.

The effects of organic pollution clearly depend on the amount of organic matter discharged into a river, and the volume of the clean water present to dilute it. In an extreme case the B.O.D. will remain high, and the oxygen level low, and there will be no recovery before further pollution occurs. Then it will only contain bacteria, sewage fungus and particularly resistant animals like *Tubifex* worms, which have haemoglobin in their bodies and which are able to exist in almost oxygen-free water. If only a little pollution is discharged into a large volume of clean water then little modification of the flora and fauna may occur. Intermediate amounts of pollution will produce intermediate results. The most commonly observed effect is a change in the types of plants and in the species of fish. A very dirty river has no fish. The most resistant fish appears to be the eel, the least the trout. In fact trout are excellent indicators of pollution; if they can exist, the water will be fit for almost any purpose.

It should be noted that we have so far been considering what I should call 'ecological pollution', with a change in the balance of plant and animal life. For man, particularly when we are considering the field of public health, there may be other considerations. Particularly in the tropics we may find water which shows none of the ecological changes characteristic of pollution but which is infested with pathogenic organisms, including the free-living stages of helminths which parasitize man and domestic animals. Here man is simply playing his part as an intermediate host in the complicated life history of the species concerned. The excrement which carried the parasites into the water has been absorbed and has helped to nourish the indigenous flora without upsetting its composition. To avoid infection with such parasites, man may alter the whole economy of the river, for instance by canalizing it so that all the flora and fauna are practically eliminated, or by using chemicals which alter the entire ecological balance.

3.2   Pollution by sewage

Today most domestic sewage in Britain is treated before discharge to inland waters. Most of the organic matter is removed, mainly by sedimen-

tation and the action of aerobic micro-organisms which act on it as the
sewage in suspension passes through filter beds. Unfortunately in many of
our cities the population has outgrown the capacity of the sewage works,
and so the treatment is incomplete. Some completely raw sewage is also
discharged. However, the situation is generally improving, and it is likely
that in a few years time uniformly high standards of what is known as
'secondary treatment', with the removal of the organic matter and the
production of effluents with very low B.O.D.s, will prevail. This will be a
great improvement, but it will not restore our rivers to their pristine
purity, and problems of unwanted algal growth in water storage reservoirs
may even get worse. This is because, in removing the organic matter, we
liberate increasing amounts of nutrient salts, particularly phosphates, into
the otherwise purified effluents. This gives rise to 'eutrophication', a
problem discussed below (p. 27).

Nevertheless there are substances in domestic sewage which, at present,
pass through the treatment plant and pollute the effluent. The most
notorious are the synthetic detergents. A few years ago many of our rivers
were disfigured by masses of white foam sometimes known as 'detergent
swans'. This was caused by the so-called 'hard' detergents which were
in general use for domestic washing and by the textile industry. As little
as 1.0 ppm of these substances caused foaming in the rivers, and reduced
the uptake of oxygen considerably. They did not appear to be particularly
toxic, and coarse fish continued to be caught by fishing through the holes
in the foam. This problem has now considerably abated, as detergents with
a slightly different chemical composition which are more easily decom-
posed by bacteria in the sewage works are substituted. However, some
industries still seem to use the resistant substances, so local pollution still
continues.

Today the main complaint regarding detergents is that they release
large amounts of phosphates which pass into the effluents (Fig. 3-2).
Attempts are being made to produce new washing powders which are low
in phosphate, but so far this has not always been successful as they have
themselves been more toxic, and have killed fish and water insects. No
doubt in time this problem will be solved. Incidentally it seems that the
so-called 'biological' washing powders, containing enzymes, do not present a serious pollution problem. They may be dangerous to the factory
workers who produce them, and cases of dermatitis among housewives
using them have been reported, but the active ingredients seem to be
broken down in the sewage works.

Domestic sewage is thus seldom acutely poisonous; its harmful effects are
the encouragement of the wrong sorts of organism in water enriched by
sewage effluent. Industry, on the other hand, may discharge highly poisonous substances. We often hear of fish killed from discharges of cyanide,
but damage from metals such as copper, zinc, lead and mercury is not
uncommon. The main trouble with some of these metals is that their
effects may be cumulative, so repeated exposures to low levels may lead

to concentration in the tissues of fish and, at perhaps higher levels, in the birds and mammals which live on them.

Organisms differ greatly in their reactions to these metals. Thus algae are particularly susceptible to copper, and a level of 0.5 ppm may be used to keep a pond or a reservoir clear. Fish can usually survive exposure to twice this level, though this margin of safety is a narrow one. The long term effects of sub-lethal exposure to many of these substances, particularly

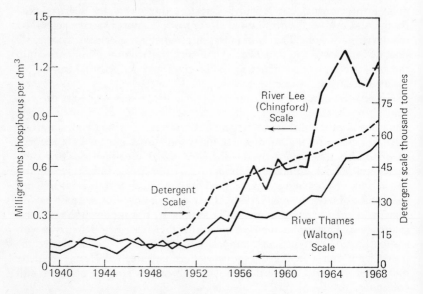

Fig. 3–2 Comparison of phosphate content of two rivers with annual consumption of detergents. (From ROYAL COMMISSION ON ENVIRONMENTAL POLLUTION, 1971)

on the inter-relation of different organisms, requires further study. A fauna, almost completely different from that found in nearby rivers with lower levels of heavy metals, is found in Wales and Cornwall in areas in which the rivers are naturally contaminated. Trout and salmon are generally the first fish to be eliminated.

3.3   Regional water authorities

In April 1974 England will be divided into 10 Regional Water Authorities, each with the responsibility for a main river basin or catchment area. These Authorities will combine the functions of extracting and purifying drinking water in their region, and in controlling sewage and waste disposal into the same rivers. It is believed that the possession of this dual function will produce cleaner rivers, for those fouling them will also have to pay for cleaning them up. There has unfortunately also been some pressure to reduce existing Common Law rights which theoretically allow riparian owners to insist that only clean water flows past their land, and to prosecute anyone polluting it. The Common Law has been successfully used, particularly by the Angler's Co-operative, to abate pollution in rivers which contain valuable fishing rights. It serves to prevent comparatively clean rivers from being polluted, but it may be impossible to identify the culprits where many outflows contribute to the filthiness of a really polluted river. The Water Authorities could in theory abolish most river pollution tomorrow, but this would mean that no water closets could be flushed by the 60 per cent of our industrial population whose local authority plant is deficient or overloaded, and mass unemployment would result from the closure of a great number of factories. Our authorities have preferred to hasten slowly, and no doubt this policy will continue.

In fact there has been a great improvement in the discharges from sewage works and industry in many parts of Britain. This improvement has been particularly noticeable in the London Thames. In the middle of the nineteenth century this became grossly polluted, and as recently as 1957 fish seemed to be totally absent, and the water was mostly anaerobic. Since then there has been a substantial improvement, and by 1968 some 41 species of fish, sometimes in small numbers, had successfully reinvaded the water. The freshwater species had penetrated downstream from the cleaner upper reaches, the marine or euryhaline species from the sea. There has been less improvement in some industrial areas, but it is probably true to say that every major river was at least a little cleaner in 1972 than it was twenty years earlier.

We seem well on the way to conquering the water pollution caused by urban sewage and industrial effluents, and it should even be possible to contain the additional burdens produced by a moderate increase in population and by some growth of industry. It is clearly going to be difficult to obtain sufficient water to keep up with the growing demand, so re-use (which is already practised to an enormous extent by industry) will grow, and there will be every inducement to see that all possible effluents are pure enough to make their re-use practicable. Already many rivers, in dry periods, consist mainly of sewage effluent, and much of our drinking water has been drunk before and purified for our use. However, as urban and industrial pollution is overcome, agriculture is giving rise to new and, possibly, more intractable problems.

3.4 Pollution by farm animal wastes

In the past, farm animals lived mostly outdoors. The return to the soil of their excreta was the main method of sustaining soil fertility. Today an increasing proportion of cattle, pigs, poultry and even of sheep are confined for all or a substantial part of their lives to buildings. Many of the farmers rearing these livestock have no arable land on which to use their manure, nor, if they have the land, do they have the labour with which to spread it. Thus much of the excreta has become a source of pollution instead of a valuable resource. Animal excrement pollutes in exactly the same way as human sewage, except that there may be more of it. Thus a cow may produce the equivalent of ten men, a pig of three, and ten hens are equivalent to one man. As in 1970 there were in Britain approximately 13 million cattle, 8 million pigs and 133 million poultry, their potential contribution to pollution was more than three times that of the human population which is already grossly overloading parts of the existing sewage machinery. Fortunately much of this muck is still used for manure, but this proportion is decreasing, and serious problems exist in parts of the country. As the number of workers on the land has decreased from not far short of a million in 1948 to only about a third of that figure in 1970, and the number is still falling, and as spreading manure is what is now called a 'labour-intensive' process, there is a considerable possibility that more and more manure will become a financial embarrassment and a source of pollution. The main hope of a reversal of this trend comes from the recent report of the Agricultural Advisory Council, which shows that in unstable soils the fall in the level of organic matter, caused partly by not using farmyard manure, is producing a serious deterioration in soil structure and a fall in the output of cereal and other crops. However, there has as yet been no sign that, at least where the soils are not unstable, there will be any substantially greater use of animal manure on the land, and the problem of pollution will probably get worse and not better, unless some additional incentives are given to use these organic manures. At the moment there is an actual financial disincentive, as inorganic fertilizers are available at uneconomically low prices because of the fertilizer subsidy; this acts as a 'negative subsidy' for organic manures which receive no such support.

Significant but decreasing amounts of untreated animal wastes are discharged directly into our rivers. This acts in the same way as untreated urban sewage, causing deoxygenation and all the ecological changes described above. Fortunately this practice, which may be legal where it was used before the Water Resources Act of 1963 or earlier legislation was enacted, is becoming less common and will eventually cease. But the rest of the animal sewage will continue to present problems. Some 'factory farmers' have retained only a few acres of land, an area considered just sufficient to absorb the animal wastes. These are regularly spread in quantities which virtually poison the soil. Under favourable circumstances there is little nuisance, and the organic matter decomposes before it is washed out. Too often, unfortunately, heavy rain washes the raw wastes into streams and rivers where the inevitable pollution and deoxygenation occurs. There is no obvious solution to this problem; the practice is clearly dangerous and is growing. On the other farms the animal wastes are run into oxygenation ditches, generally harmlessly, as the organic matter eventually decomposes, but these ditches may overflow in heavy rain and again pollute the rivers. In some instances farmers dispose of their wastes into local authority sewage works, where these have surplus capacity and the farmers can afford the substantial charges. Other farmers have installed their own sewage works, perhaps at a capital cost of £20 000. Many who are less fortunate have gone bankrupt, when the additional burden of waste disposal has made their whole farming operation uneconomic.

Today an increasing amount of silage is made, to conserve grass and other crops for use indoors and in winter. Nutritionally this is an excellent practice, but unfortunately silage also produces an effluent which has an immediate effect in polluting water into which it escapes. As more and more silage continues to be made, such accidents may become more frequent.

3.5 Eutrophication

Where animal wastes are disposed of into ordinary sewage plants, the organic matter is retained, but a substantial amount of the nutrient salts is passed out into the effluent, and contributes to the problem of eutrophication, in the same way as does urban sewage. Eutrophication, in which water becomes more heavily loaded with nutrient salts, is a natural process. Upland rivers and lakes are usually 'oligotrophic'; they contain few nutrients in solution. They remain clear, as plant growth is not encouraged, and fish like trout which feed largely on insects and other forms of animal life nourished outside the water flourish. Such waters naturally become eutrophic as they flow into the lowlands, being enriched by salts leached out of the land through which they flow. A young lake is oligotrophic, as it gets older nutrients accumulate in it and it becomes eutrophic. Under natural conditions eutrophication gives rise to a balance, detected in the difference between the flora and fauna of a lowland river or lake from those of the uplands. Trouble occurs because, where eutrophication is caused by man, the time scale is so reduced that this balance is upset.

Sometimes, as in fish farming, eutrophication is deliberate. The intention is to maximize production, with an interaction between the plant and the animal life, and a surplus production of fish which can be cropped. Considerable skill is needed to preserve a balance. Unintentional eutrophication seldom does so.

As has already been shown, nutrient salts are deliberately added to our rivers in sewage effluents and as animal wastes, and accidentally when the excrements of farm animals are not properly disposed of. Modern arable farming also contributes. It has been shown that, in some but not all of our rivers, levels of nitrogen have doubled between 1958 and 1968. Most of this increase has come from arable land. Cereal crops in Britain have given ever increasing yields during the last 25 years, except in the limited areas where the soil structure has deteriorated as mentioned above, and even here the trouble has only proved serious in unusually wet years like 1968. This increase has been caused largely by the use of varieties of cereal which respond to high levels of chemical fertilizers. Thus the average yield has increased from below 2000 kg/ha in 1949 to over 3000 kg/ha in 1970. During this period the amount of nitrogenous fertilizer used in Britain has increased greatly, from 300 000 tonnes in 1957 to 650 000 in 1968. Unfortunately only about half of the nitrogen applied to the land is taken up by the crop; the rest is lost, and some contributes to eutrophication. The equation, however, is not a simple one. Nitrogen is lost from all soil regardless of whether or not it is fertilized with organic or inorganic manures, or is synthesized by nitrogen-fixing bacteria. The exact way in which losses occur is still uncertain. Some, but certainly not all, is leached out into the drainage water, and some is turned into atmospheric nitrogen by denitrifying bacteria. When farmyard manure was applied in bulk to bare soil in the autumn, the greater part of its nitrogen was lost during the winter. This had little effect on rivers, as the temperature was too low for massive algal growth and the salts passed harmlessly, but wastefully, to the sea. It is often assumed that salts in inorganic fertilizers will be more readily and rapidly lost than those in farmyard and other organic manures, and there is evidence that this may be the case especially when the soil structure deteriorates, but some inorganic fertilizers, used at the right time, are efficiently converted into plant tissues. Thus when grass is given a top dressing in early spring to stimulate its growth, the salts are usually efficiently and almost totally taken up by the mass of fibrous roots under the turf. However, this process can also give rise to a particularly serious form of pollution. If the fertilizer is applied to saturated soil after a wet winter and then there is heavy rain before the material is incorporated in the soil, it may be washed off and give very high levels, as much as 30 parts of nitrogen per million of water, in the rivers. Where these rivers are used as water supplies this may be dangerous. Purification of river water removes much of the pollution, but soluble nitrates usually remain. They are of little danger to adults, but young animals including human babies have a different flora in the gut, and this may turn the nitrates to nitrites which are poisonous, acting by combining with the haemoglobin in the blood to form methaemoglobin which does not carry oxygen from the lungs to the tissues. Fatal methaemoglobinanaemia in babies has occurred in Europe and America, caused by drinking water containing too much nitrate. In parts of East Anglia, wells in rural districts have been found with dangerously high nitrate levels and one infant's death has been attributed to this cause. Bottled water is imported for drinking in a number of areas.

To affect the aquatic vegetation, water must contain the various salts necessary for growth. The different components causing eutrophication come from various sources. Nitrates are leached mainly from agricultural land, but little phosphate, even when liberally applied in chemical manures, is lost as it is usually bound firmly in the soil. Phosphate levels have increased spectacularly since 1952, and these are believed to come mainly from detergents. There is a close correlation between the increased use of detergents and the rising phosphate levels in several British rivers. There is some argument as to whether phosphate or nitrate pollution has the most serious effects. They probably both make their characteristic contribution.

Eutrophication affects the vegetation of running water, but its greatest effects are seen when this water is impounded in a reservoir or where the river runs into a lake. The most obvious result is an algal bloom. Many species of algae are involved, sometimes unicellular forms (*Monodus* sp.) pullulate to turn the water into something like pea soup, at other times the whole surface is covered with a mat of blanket weed (*Cladophora*). Algae give considerable trouble if the water is to be used for town supplies, as they choke the purifying plant. They also induce deoxygenation. This sometimes causes surprise, as they are green plants which produce oxygen during photosynthesis, so they might be expected to have the reverse effect. They do of course liberate oxygen, but with blanket weed this is at the surface and much is lost, while the lower layers of the water are in shadow where plants cannot flourish. Finally the algae die and decompose, and this removes oxygen as does the decay of any other form of organic matter.

Algal blooms do not occur in pure water containing few salts, but they have been observed when NO_3—N levels are as low as 0.3 ppm and PO_4—P has been as low as 0.01 ppm. Such and even higher levels occur in many waters where blooms are infrequent, though algal growth is commonest where eutrophication is most intense. We do not always know what triggers off an algal bloom. In some cases grazing crustacea, including *Daphnia*, may keep the algae in check, and if these are killed by a spill of insecticide, the bloom takes place, uncontrolled. Sometimes where nitrate is low and phosphate high blue-green algae (e.g. *Anaboena* sp.), which can taint the water, are found in some abundance, for these plants are able to synthesize nitrates from atmospheric nitrogen. The whole question of the harmful effects of eutrophication and the conditions which encourage unwanted algal blooms, requires further study. Sometimes there is a balance, and algae though growing rapidly are kept at bay by being grazed by *Daphnia* and other crustacea. This cannot always be relied upon to continue so every effort should be made to prevent nutrient levels rising suddenly as a result of pollution.

Some waters are rendered eutrophic when polluted by birds. Reservoirs and lakes are favourite roosting places for gulls, which sleep overnight in their thousands on the surface of the water. Unfortunately they also excrete into the water, increase its nutrient status, and precipitate algal blooms, with just as serious results as when man is the culprit.

Life can only exist over a comparatively narrow range of temperature. Warm blooded animals, that is mammals and birds, have developed control mechanisms which maintain nearly constant body temperatures independent of those of their surroundings, but even they may be killed by extremes of cold and of heat. The rest of the animal kingdom, and all plants, are profoundly affected by changes of temperature. Each organism has a cold death point, and a rather narrow range of temperature where life can proceed. As a rule within this zone metabolism is slowest at the cold end and increases as it gets warmer. Above this zone are higher temperatures where exposure is unfavourable and, eventually, fatal. Laboratory experiments to determine the effects of high temperatures have only a limited ecological significance. Thus a fish may live apparently normally in a fish tank, but may never be found in a river heated to the same temperature. The whole balance of animal and plant life in a habitat may be upset by a comparatively small change in temperature, so that an organism which appears to live normally in isolation may be eliminated by other species which find the new temperatures even more favourable.

If some activity increases the temperature so that man, plants or animals are harmed, this may clearly be considered to amount to 'thermal pollution'. In North America there are several widely reported cases of rivers being heated almost to boiling point so that they are completely lifeless. Other cases of water at 50 °C or above contain no fish and practically no invertebrates, though a few thermophyllic bacteria may flourish. Such cases are not widespread. As a rule industry tries not to damage the environment. So far no really spectacular thermal pollution has occurred in Britain, but increasing industrialization is having its temperature effects and these need to be examined. The most notable are where water is used to cool machinery, particularly electrical power stations.

4.1 Dissolved oxygen levels

Biologists are particularly worried lest heating water may cause deoxygenation. If well oxygenated water is warmed, the level of oxygen in it is reduced. This is because, as the temperature rises, the solubility of oxygen in water decreases. Thus at 5 °C a litre of water can take up 9 cm³ of O_2, while at 20 °C only 6 cm³ will dissolve in the same volume. This imposes difficulties on an aquatic organism, for a rise of temperature from 5 °C to 20 °C is likely to increase its rate of metabolism, and therefore its rate of respiration (and its need for oxygen) by a factor of about four. Of course if the water which is warmed is, at the lower temperature, only two thirds saturated, warming will not expel oxygen until a temperature above 25 °C is reached. Thus heating polluted water may have less effect on the oxygen content than a similar rise in temperature under purer conditions. The fauna of unpolluted water is also more easily destroyed by environmental change so, if it is used for cooling as industry spreads to rural areas, we must be on the look out for harmful effects.

In Britain the electrical industry is the greatest user of cooling water. The scientists in the Central Electricity Generating Board research laboratories have made studies of the effects of power stations on the rivers which they use. Thus at Drakelow on the river Trent some 862 150 cubic metres of effluent from the cooling towers was returned to the river in summer, and 645 080 cubic metres in winter. Less water was required in winter as it started off colder, and so could absorb more heat. The average temperature of the intake over 24 hours in summer was 18.2 °C and that of the outlet was 27.4 °C. In winter the corresponding figures were 6.6 °C and 20.4 °C. Rises of this magnitude obviously had a substantial local effect, and temperature rises, to 22.4 °C in summer and 12.4 °C in winter, were measured 1.6 km downstream so a large area of the river was affected.

Contrary to common belief, this warming of the effluent has not caused deoxygenation, in fact the oxygen level in rivers has been improved. The Trent is a polluted river, with the oxygen level falling as low as 17 per cent of saturation, and seldom rising above 50 per cent. The water used in cooling is saturated or even, to some extent, supersaturated. As it is warmed, the amount of oxygen taken up is limited, but nevertheless at Drakelow between four million tonnes of oxygen, in summer, and two million tonnes in winter, are added to the river every day.

4.2 Ecological effects

These changes in temperature and oxygen content of the Trent must obviously have ecological effects. Tubificid worms, able to live in the polluted water, appeared to breed more successfully below than above the power station, and the breeding season was extended, so that it reached its peak in October, when breeding in the cooler water above the station did not take place. In the less polluted river Severn near the Ironbridge power station similar temperature changes occur. Their effects on the distribution and life histories of several larval stoneflies (Plecoptera) and mayflies (Ephemeroptera) have been investigated. Very little effect of the increased temperature was observed, though several species continued to grow only in the warmer waters during the winter. So far no significant changes in the balance of various species has been discovered.

Again referring to America, where greater temperature changes have been produced, more obvious results have been observed. These have usually been, from the anthropocentric view, for the worse. Salmon and trout have been replaced by coarse fish; the total protein production has been increased, but not the quality of the environment. There has generally been a loss of species, and a reduction in diversity. Such changes in Britain may be expected if greater temperature changes are induced by an increase in the use of river waters for cooling. But so far, this would not seem to be a serious cause of what should strictly be called 'thermal pollution'.

Nuclear power stations are mostly on the coast, and are cooled by sea water. This does not seem to cause much damage. The water in the vicinity is certainly warmed, and there are plans for farming oysters and mussels, though these have received limited support for fear that the molluscs may concentrate radioactive materials in the effluents. The main trouble has occurred within the cooling system, where young stages of mussels may pass through the grids and settle down inside the cooling tubes. They grow unusually fast in the warm water and interfere with the circulation.

Obviously it would be prudent to try to use all this surplus heat, perhaps for fish farming under controlled conditions. If this is done, further pollution will perhaps be avoided. But the main worry will continue to be that warming waters which are so far unpolluted may result in ecological degradation of a rather subtle and long term variety. Gross thermal pollution, with the sudden death of fish from heat or deoxygenation, will be much easier to detect, and good long term planning should prevent it from happening.

4.3   Global effects

To many people, thermal pollution signifies global changes of climate with adverse effects on man and the economy of the globe. Most of the energy produced by industry is eventually wasted as heat, and this must raise the temperature at least in the vicinity of its production. The question is whether such a rise is significant. We know that the climate of a large conurbation differs from that of the surrounding country; the average temperature may be as much as two degrees higher. This affects plant and animal growth, though results are not spectacular, perhaps because in the past smoke has reduced the light intensity and caused opposite effects.

At the moment the heat liberated by all power sources is an insignificant fraction of that which reaches earth from the sun. Nevertheless as our energy requirements are doubling every eight or so years, an exponential increase of this kind could eventually reach, and even surpass, the heating capacity of the sun's radiation received by this globe. Any argument based on exponential growth is liable to lead to a nonsense: thus for the world's industry to equal the heat out-put of the sun's radiation, the whole of the land surface would need to be covered with generating stations. The curve for power requirement is likely soon to flatten off, if for no other reason than that fossil fuels are becoming exhausted and nuclear power production has its limitations. It is unlikely therefore that man-made heat will alter our climate, except locally. These effects hardly deserve the name of pollution.

The possibility that pollution will alter the climate of the whole globe by affecting the proportion of the sun's energy reaching it requires more serious consideration, though, as mentioned in Chapter 1, possible changes have probably been overestimated. But pollution from high flying aircraft does produce particles in the stratosphere and many pollutants including ammonium sulphate produced from SO_2 may induce cloud formation, and these could affect the 'albedo', or reflectivity, and so reduce the amount of radiation reaching the world's surface. Changes in atmospheric composition, particularly of CO_2, could trap differing amounts of the energy which reaches the world. So pollution could cause the world to become cooler or hotter. It is most important that these changes, and their effects, should be widely monitored, even though it seems at present likely that natural fluctuations of temperature will be greater than those which man, with his present technology, can produce. But technology will develop, and there will be conscious efforts to 'improve' the world's climate. There is a risk that improvements in one region might be harmful in another.

Radiation

5.1 Harmful effects of radiation

No one doubts that radiation can be harmful to life. In really high doses, it causes immediate death. A slight decrease in this intensity may not be immediately fatal but will cause obvious burns and other symptoms, and death is usually not long delayed. A somewhat lower exposure may cause no immediate symptoms, but within a few days 'radiation sickness' manifests itself, and death commonly occurs; those who recover may have their health permanently impaired. Even lower exposures may have no recognizable effect on an adult, but may cause abnormalities in a mammalian embryo *in utero*. A considerably lower intensity may cause no detectable effect in a single dose, but if there are repeated exposures harmful effects, commonly in the form of cancer, will result. It seems that, within limits, the effects of small exposures are cumulative, so that every small dose of radiation means that the risk of damage from further exposure is proportionately greater. Finally the very smallest dose may cause no recognizable effects in the victim, but the germ cells may be affected, so that mutations occur and the results of the exposure may only appear in succeeding generations.

Pollution by radiation can be said to occur when levels of radiation are such that they have harmful effects. It is difficult, if not impossible, to fix a low level where the pollution can be assumed to be completely harmless. Radiation consists essentially of electro-magnetic waves or various types of sub-atomic particles which cause damage when they enter living tissues. It seems that even the lowest level of radiation will do some damage, if the appropriate target is struck. It is most improbable that a low level will cause death or radiation sickness, for far too few cells will receive any damage to produce detectable reactions. But in theory at least a single electron set free by an incredibly minute level of radiation could damage a vital cell which, by division, could lead eventually to the formation of a malignant cancer. It could also affect a zone of a chromosome in the testis or ovary such that a hereditable and harmful mutation occurred, only detectable if it were recessive, after several generations. But the probability of this type of damage arising from such a dose is remote.

We are all subjected to a background level mainly of naturally occurring radiation. The level is in the region of o.1 rads (the rad being a unit measuring the dose of radiation) absorbed by the body. The origin of this background radiation is shown in Table 1.

It will be seen that, so far, man's contribution is very small. We obviously cannot avoid this background radiation, and some people have assumed that, because the human race has evolved under these conditions, it is

Table 1

Source	Annual dose in rads
Natural Background	About 0.1000
made up of:	
Cosmic rays	0.0250
From rocks	0.0500
From within the body	0.0250
Fall out	0.0013
Waste disposal	0.0003 to
(from nuclear power)	0.0030

totally harmless. Clearly we can exist under these conditions, but they, like naturally occurring poisons, could well be as harmful as man-made radiation. For that reason attempts are made to reduce additions to as low a figure as is practicably possible.

5.2 Permissible levels of radiation

Permissible radiation levels are often related to the natural background levels. For food which might be contaminated from a controllable source a figure of one thirtieth of the background is sometimes stipulated as the highest permissible dose to any consumer. For radiologists and radiographers, who in the early days of X-rays were among the victims of overexposure, the 'maximum permissible dose' has been reduced four times since 1931 (Fig. 5–1), in the light of experience, and the level now fixed is that below which radiation could not be effectively used in medicine, industry and research. This does not mean that no one is harmed; it means that the risks are considered to be sufficiently small in comparison with the benefits to the community.

In fact we have two standards in industry and medicine. The Ministry of Labour in its 'Code of Practice' lists what are called 'designated persons' whose work exposes them to particular risk; they are sometimes allowed to experience a higher level of radiation than is generally acceptable, but it is usual to subject them, subsequently, to stringent safeguards so that a total dose of a dangerous level is not reached. Older men, with less risk of subsequent exposure, are often permitted to receive more radiation than their younger colleagues.

It should be stressed that though we have no firm proof of damage to individual men from these very low levels of radiation, good circumstantial evidence exists, for instance from the higher levels of cancer in radiologists than in the general population, and from the fact that, in America, while the average age at death of the general population is 65.6 years, that of radiologists is only 60.5 years. The exact way that this exposure to radiation has affected the bodies of those concerned is still not fully understood.

Though those receiving specially high doses of radiation are clearly at risk, it is more difficult to assess the possible danger of existing levels of world pollution. It may sound very alarming when it is said that the level of some radioactive substance has increased, perhaps even by a factor of over a thousand, because of testing atomic weapons, but if even after this increase the absolute level of radiation produced is still only a tiny fraction of the background level then the risks may not be very great. Thus the radioactive isotope Strontium[90] has been produced because of the testing of nuclear weapons. Strontium is taken up by plants, and it was shown that in Wales the levels of Strontium[90] increased from virtually nil in

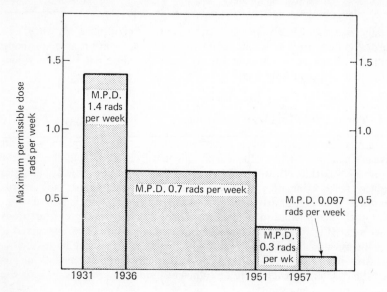

Fig. 5–1   Chart illustrating the reduction in the Maximum Permissible Dose (M.P.D.) of radiation over the years. (From PIRIE, 1958)

1954 to 2500 picocuries (or micromicrocuries, see §5.4, p. 38) in 1956 (Fig. 5–2). This apparently enormous increase has in fact only increased the background radiation level by a fraction of one per cent, so its biological effect may not be great.

5.3   Biological effects of radiation

This, however, is not the whole story. The danger from any radioactive substances depends on where they are located. Strontium[90] is often spoken of, somewhat emotionally, as a 'bone seeker'. This is intended to indicate that it may be concentrated by man from his food or drink, and stored in his bones. There is not any special concentration of radioactive substances.

Bones normally contain much calcium and a substantial fraction of strontium. Man and animals naturally concentrate these elements into their bones and, if radioactive strontium is present, it is also concentrated together with the normal element. The danger from Strontium[90] is particularly great because it is concentrated adjacent to the bone marrow. The marrow is active in blood production, and radiation may cause a type of cancer which manifests itself in the fatal disease leukaemia. Since weapon testing began, there has been a slight world-wide increase in the incidence of leukaemia, and many scientists correlate this with radiation and in particular with this increase in Strontium[90]. The risk to any one individual is still so small as to be almost unmeasurable, but nevertheless

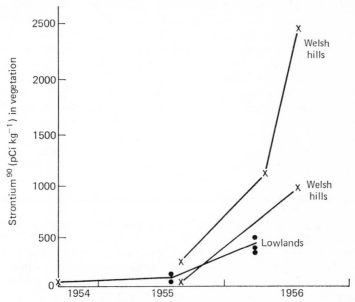

Fig. 5–2   Strontium[90] in vegetation in Britain, 1954–56 (picocuries Strontium[90]/kilo). (From PIRIE, 1958)

several thousand cases may well have developed among the whole of the world's population. This and the risks from other isotopes is sufficient to make the complete cessation of weapon testing desirable. Since above-ground tests have been restricted, world radiation levels have in fact decreased, but could easily increase again if the restrictive agreements broke down or if more countries evaded them.

If we survive nuclear testing and the almost certain extinction which would result from nuclear war, we will still be subject to man-made radiation from medical and industrial sources. As already mentioned, we are becoming increasingly stringent in our control of X-rays and of radiation used in medicine, and have reached a situation where any further reduction would mean abandoning many useful techniques. It is generally agreed that less damage is likely to the health of mankind from the present levels of radiation than would inevitably result from abandoning these techniques for medical diagnosis and treatment.

5.4   Radioactive wastes

The use of nuclear power is increasing. So far very rigorous safeguards have been enforced, so that though accidents have happened and some radiation has 'escaped' from power stations, little obvious damage has resulted. We are producing an increasingly large amount of radioactive waste material. Some is discharged in the air and into the sea, but levels have been set to avoid serious risks to the general public. Thus at Windscale the standard is set by the level of radioactivity in lobsters within range of the effluent. The effluent must not contain enough radioactivity to raise levels in lobsters, which are efficient concentrators, high enough to endanger someone eating a large specimen daily throughout the year. Other radioactive waste cannot be disposed of so easily, and has been stored in deep mines or dumped in concrete containers at sea. Much of the radiation will cease as the radioactive isotopes decay, but some emit powerful amounts for very many years. Already waste disposal is getting more difficult. That dumped deep in the earth may escape, particularly when earth tremors occur. There is grave concern about the safety of dumping at sea.

The problem is a growing one. Our fossil fuels are being used up, so nuclear power will be used to produce a greater proportion of the rapidly expanding power production in the world. Today in the United States some 150 cubic metres of the waste produced annually contain over a thousand megacuries of radioactivity. A megacurie is the activity of a million grams of radium; only some 10 000 g of radium itself have in fact been isolated for all medical uses in hospitals in all countries. The size of the problem may be indicated when it is realized that the levels of radioactivity in vegetation etc. polluted by nuclear fallout are measured in picocuries (or micromicrocuries), and that one megacurie is 10^{18} picocuries. By the year 2000 the annual level of U.S. radioactive waste may reach nearly a million megacuries, and energy requirements will no doubt increase after that date, so disposal will become an even more serious problem.

The larger the number of power stations, the greater is the risk of an accident, with the release of a dangerous amount of radiation. It seems likely that with an enormous number of these installations in all countries, it will be difficult to ensure that all observe the type of precaution at present in force in the few existing stations, particularly with regard to the disposal of radioactive waste. Thus though nuclear power has, so far, done little harm, in the long run it may well be the most dangerous form of man-made pollution.

5·5 Mutations

As already mentioned, there is much concern lest radiation should cause harmful mutations, particularly in man. However, we have little evidence of the results of low levels of radiation in actual practice. There is no doubt that moderate intensities of radiation can cause mutations. This has been demonstrated in the laboratory with the fruit fly *Drosophila*, with mice, and with various crop plants. The mutations produced are almost all harmful, producing deformed organisms. Lethal genes, causing the embryo to die, have also been produced. But the level of radiation has usually been higher than would be tolerated in the general environment for health reasons. In theory much lower levels could be effective, but this is difficult to prove by animal experiments as such huge numbers of animals would be required. Thus we have little hard evidence of the mutagenic effects of man-made radiation outside the laboratory. The background radiation may have been an important factor in producing the mutations on which the whole process of evolution depends, but this again is a speculation.

One experiment suggests that radiation-induced mutations may not be as frequent as is often imagined. In the Brookhaven Forest in the United States a fascinating experiment has been in progress for the last ten years. A massive source of radiation switched on for some 20 hours of every day has killed all life in its vicinity. Further away more selective damage has been done, and valuable results on the susceptibility of different plants and animals have been obtained. Further still from the source the radiation, though far above the levels tolerated for human exposure, causes no obvious damage. Nor has it apparently produced any obvious mutations in the plants and animals subject to these different levels of exposure. This may indicate that the risks of producing mutations may have been over-estimated. However, as quite low levels of radiation can obviously cause damage to man and to other organisms when this is studied epidemiologically in a large population, even where no obvious damage can be observed in an individual, it is obviously right to try to keep radiation as low as is practicable.

Pollution of the Sea 6

The oceans cover 56×10^6 sq. km and contain 1420×10^{15} cubic metres of water. This is an awful lot of water! It corresponds to nearly 5×10^8 m^3 for every inhabitant of the earth; it is not surprising, therefore, that we often assume that the sea can absorb unharmed all the wastes we like to discharge into it. Unfortunately, however, pollution of at least some of the seas does occur, largely because noxious substances do not become equally mixed but remain concentrated in limited areas, or may even be reconcentrated by living organisms after apparently safe dilutions have been achieved.

6.1 Oil pollution

Oil is the most obvious pollutant of the ocean. The danger from this source received the greatest publicity after the tanker the Torrey Canyon went aground on rocks off the coast of Cornwall on 18 March 1967, and when more than half its cargo of nearly 120 000 tonnes of oil escaped. Even before that there had been many other serious, if smaller, incidents when tankers had been wrecked or damaged, when ships had deliberately discharged waste oil, and when tanks had been washed out prior to reloading. Oil has caused damage in the open ocean and in estuaries, and even where spills have occurred in rivers well inland.

The oil industry is so huge that it is hardly surprising that accidents occur. Tankers now carry annually some 8×10^8 tonnes of crude oil around the world, and it is said that several million tonnes of waste oil are today floating on the oceans, often harmlessly, but still a potential danger if washed ashore or into a region where birds were diving to catch fish. Until recently it was policy to wash tanks out at sea, adding several million more tonnes to the ocean's pollution. Recently this has decreased as the 'load on top' system has been adopted by most of the leading oil companies. In this the oily waste at the bottom of all tanks after washing is collected in one tank. Here the oil floats to the top of the water which can then be safely pumped out from underneath. The oil is retained and the next load is added to the tank without further loss.

Oil is a serious pollutant in its own right. The effect on sea birds is well known. Their feathers become clogged, and they cannot fly. In attempts to clear the plumage, the birds preen themselves, and in doing so imbibe sufficient oil to poison themselves. The oil interferes with the insulation provided by the feathers, and the birds die of cold or their susceptibility to pneumonia increases. The Torrey Canyon incident may have killed

100 000 birds of different species; guillemots were the most severely affected.

The results of oil pollution on birds are distressing, and there is no doubt that they are harmful to many individuals. The effects on the populations of the different species are more difficult to assess. Guillemots and other auks are now less common on the coasts of much of Southern Britain than they were 30 years ago, and this may be the result of oil pollution. However it must be acknowledged that no species is in danger solely as the result of oil, and the ability of many to increase after there have been many deaths is remarkable. So though oil pollution does seriously affect many birds, its result on a world scale may not be as serious as is sometimes imagined.

Oil affects many other marine organisms. Where rocks are thickly covered with oil seaweed dies, and many molluscs and crustaceans are also killed. But the remarkable thing is that the recovery at least from a moderate amount of oil is remarkably rapid and complete. Bacteria 'live on' the oil and break it down. Within a year of quite heavy oiling, rocks and beaches have been almost completely restored, and their flora and fauna have regenerated. The most serious damage has been done by the attempts to combat the oil pollution itself.

The greatest economic damage done by oil is when it lands on the beaches of sea-side resorts. For years tarry patches have soiled the skin and the exiguous bathing attire of holiday makers and dirtied the feet of paddling children. The problem is an unpleasant one even when the oil is present in small amounts. A massive spill can cover the whole surface of the sand and drive all holiday makers away. Various detergents have been used to emulsify and disperse the oil. They have been reasonably successful in this task but unfortunately they have proved to be much more toxic than the oil itself to many marine organisms.

The toxicity of detergents was realized before the Torrey Canyon incident, for it had been observed that where part of an oiled rock had been treated, recovery of the flora and fauna was slower than where the oil was left untouched. However, the possible economic damage of the Torrey Canyon spill was such that some 3×10^6 dm^3 of various detergents were used to clear and protect the Cornish beaches, notwithstanding the reservations of the biologists. As the substances used (nonyl-phenol polyoxyethylene condensates) are toxic to crustaceans at levels well below one part per million, it is not surprising that considerable damage was done. The surprising thing is that it was not much more serious. Although many species of molluscs, crustacea and fish suffered heavy casualties, and large areas of rock were denuded of their seaweeds, yet recovery has been remarkably rapid and within two years was virtually complete. No permanent losses of any species of animal or plant have been reported.

The oil from the Torrey Canyon would have done more damage in British waters had not the wind changed before all the oil had been blown ashore and diverted it to the coasts of France. Here considerable local damage was done to fish and oysters, but little detergent was used. Instead the water was covered with sawdust and chalk. Oil which was sunk in this way did some damage to the organisms on the bottom, but again it was broken down fairly quickly. More oil was washed ashore and trapped in the salt marshes. Here again local and transitory damage occurred, though examination in 1968 suggested that some species were more vigorous, their substrate having been 'fertilized' by the oil.

These studies should not be allowed to suggest that oil pollution is unimportant. Had the whole of the Torrey Canyon's load gone ashore rapidly in Cornwall, as it might have done with slightly different weather conditions, the results would have undoubtedly been more serious and longlasting. It is now generally agreed that detergents should only be used on beaches frequented by holidaymakers, as natural processes remove oil fairly rapidly unless the concentration is too high.

6.2 Sewage and poisonous wastes

A large amount of raw sewage is discharged into the sea. Where long outfall pipes are used, so that there is little risk of fouling beaches, this system is reasonably unobjectionable. The organic matter breaks down and the nutrients are recycled. However, the discharge is often too near to the land, and the shore may be fouled. This is unaesthetic, and there must be some risk to public health, but the risk is less than is often imagined. A thorough investigation in towns in Britain where such pollution of the beaches was common showed that, even among those who swam among the sewage, there was little evidence of infection from pathogenic bacteria or viruses.

Other organic wastes are dumped into the sea. Vegetable wastes from freezing plants in Lincolnshire discharge considerable quantities, but the results are comparatively small. Reasonable mixing seems to occur, though the local enrichment may promote the growth of mussels and some seaweeds.

Today the main concern is that the ocean may be being polluted by poisons which do not break down as rapidly as does sewage or other organic materials. There are reports of discharges of heavy metals, including mercury and lead, of industrial chemicals such as polychlor biphenyls, and of organochlorine insecticides. These last are dealt with in Chapter 7.

There is little control of the dumping of poisonous waste materials outside territorial waters, and no international agreements on this subject. Considerable industry exists to get rid of embarrassing wastes. In many cases this is done in a responsible manner, fishing grounds are avoided and durable containers likely to prevent leakage at least in the immediate future are used. Other dumping, however, is less well controlled, and local pollution has occurred.

As mentioned above, any substance completely mixed throughout the oceans would be diluted to a harmless level. Even if a million tonnes were discharged, the level throughout the oceans would be less than one part in a million million, and we know of no pollutant that is toxic at that level, or that could be concentrated by any known organism from such a dilution.

Where mercury used in wood pulp manufacture has been discharged into estuaries, levels have been high enough to damage molluscs and fish, which have concentrated the mercury to a level where men eating these animals have been fatally poisoned. There is clearly a considerable risk from these discharges into estuaries or to enclosed waters like the Baltic. Local damage to fisheries in shallow areas like the North Sea may occur before sufficient dilution has occurred. A recent scare suggested that Tuna fish was seriously contaminated by mercury discharged into the ocean by man. It now appears that the mercury occurred naturally in the ocean, and that man-made additions were negligible in comparison with the amount of the metal already present. The Tuna were able to concentrate the naturally occurring metal. Though 'natural', this could be as toxic as if it had been introduced by man. Fortunately the mercury levels in Tuna, whatever their origin, do not seem high enough to endanger the eater.

It has been suggested that lead also may be dangerous, if concentrations near to the coast are high enough. This could be derived from the tetra-ethyl lead in petrol. Local pollution may occur in estuaries and affect isolated animal populations. However, there is little evidence that the alarmist stories that the whole fauna and flora of the ocean may be so seriously polluted by metals and organic poisons that it is being destroyed are true. Nevertheless it would be prudent for an international monitoring scheme to be organized, to measure changing levels of persistent pollutants in different parts of the ocean, and to investigate changes in animal and plant populations.

Pesticides

The pollutants so far considered have been substances produced by man or his industries, which have caused unintentional damage to the environment. Their control has been achieved by taking greater care over their disposal and of their dispersal. Pesticides, on the other hand, are poisonous substances deliberately disseminated in order to exploit their toxic properties; they become pollutants when they reach the wrong targets.

Pests are organisms which man considers to be harmful, and pesticides are the substances used to control them. I shall include herbicides, used to kill weeds, fungicides, which control pathogenic fungi, and insecticides, developed for use against insects, as being the main classes of pesticides.

A pesticide is normally used against a particular organism. Ideally it should poison it, but be otherwise harmless. Although there are many chemicals with a remarkable degree of selectivity, so that the chosen pest is destroyed by a lower exposure than that required to damage other plants or animals, nevertheless complete selectivity is virtually impossible. This means that there is always the risk that pesticides will cause damage to man or to other, non-target, organisms.

Some pesticides are acutely poisonous but unstable substances. They can cause serious damage over a restricted area, but not long term pollution. Other pesticides may be less acutely poisonous, but they may be much more persistent, and they may thus continue to have ecological effects for very long periods. They may be transported over long distances, and cause damage far from the original site of application. Also a persistent poison may reappear where least expected. Although it may have been diluted down to a harmless level, it may be reconcentrated in some biological system up to levels where serious damage can be demonstrated once more.

7.1 Herbicides

Herbicides seldom cause serious environmental pollution, though they are used more widely and in greater quantities than any other pesticides. In Britain the 'hormone herbicides', the phenoxyacetic acids, are the main means of controlling weeds in cereal crops. MCPA (4-chloro-2-methyl-phenoxyacetic acid) is the most popular substance. Properly used on a cereal crop, the broadleaved weeds are eliminated with very little in the way of side effects. MCPA is not a direct 'killer'; it disorganizes the growth of the weeds which are then removed by drought and competition. The cereal seedlings are usually unharmed, though overdosing or spraying at the wrong stage of growth causes some upset. The final results are similar to the laborious weeding which preceded the use of these chemicals. There are some risks when a spray is applied carelessly or in windy weather, when the MCPA drifts onto other areas. Little in the way of 'run off' generally occurs, though spraying into ditches or ponds may kill aquatic and emergent plants. The effects on the soil flora and fauna are slight; the most noticeable is that bacteria which take part in the breakdown of the herbicide increase in plots which are repeatedly sprayed. Otherwise worms and mites in the soil seem to be unaffected. The chemical is not persistent and so long-term contamination does not occur.

Another related herbicide, 2,4,5-T (2,4,5-trichlorophenoxyacetic acid) has enjoyed some notoriety because of its use as a defoliant in Viet Nam. Here the herbicide has been sprayed, mainly from the air, at a far greater concentration than has ever been used in agricultural practice. Also the samples used have been found often to contain a high proportion of a very toxic impurity, Dioxin (2,3,7,8,-tetrachloro-dibenzo-para-dioxin) which, at least in the laboratory, is teratogenic (i.e. if applied to pregnant animals, the young may be born deformed). It seems that, in Viet Nam, long term ecological damage has been done, with the destruction of large areas of mangrove swamps and the death of forest trees. It is suggested that humans have also been harmed, and babies born deformed. However, with careful use of preparations which are not heavily contaminated with dioxin, 2,4,5,-T can be a useful chemical particularly for the control of scrub. I would think that it would be better not to use it as an aerial spray, where control of the quantity used and of the target is difficult, but otherwise I do not think that it can be thought of as a dangerous pollutant.

Other widely used herbicides are the persistent and long acting Substituted Ureas (e.g. Monuron) and Substituted Triazines (e.g. Simazine), and the quick-acting but non-persistent Bipyridylium compounds (e.g. Diquat and Paraquat).

Simazine is used in high dosage as a 'total' weedkiller on railway tracks and similar places. It acts slowly, direct spraying on plants often seeming harmless. It usually acts through the soil, from which it is not easily leached. Simazine is also used to prevent weed growth among crops, e.g. maize and asparagus, which are not susceptible to it. Very little accidental damage has ever been reported when these chemicals have been used carefully, and they are unlikely to act as pollutants.

Paraquat gives results rather similar to a flame gun, killing the above-ground parts of most plants. There is some translocation through the plant, and grasses are rather susceptible, though deep rooted thistles and convolvulus seem to recover with increased vigour. Paraquat is normally immobilized almost immediately by absorption onto the soil particles, from which it is not released and is then broken down by bacterial decomposition. If therefore it is carefully applied to its target, there is practically no risk of pollution outside that area. One great advantage of Paraquat, which is used to control weeds when pasture is resown, is that the soil fauna, particularly the earthworms, survive its application far better than

they survive normal ploughing and cultivation. However, this chemical
has disadvantages and dangers when wrongly used. First it is very poisonous to man if accidentally drunk in undiluted form, in fact it is the only
pesticide in use in Britain today which has caused deaths in recent years.
Its action is particularly unpleasant, causing asphyxiation by damaging the
lungs which are choked by the proliferation of the epithelial tissues. Eye
irritation is also caused by exposure to low volume (and therefore concentrated) spray. Secondly the chemical is much less rapidly immobilized
in wet vegetation, and when sprayed onto weedy stubble after rain wild
animals, particularly hares (which are much more susceptible than
rabbits or cattle) have often been killed. But with proper and careful use,
Paraquat should not be an environmental pollutant.

7.2 Fungicides

Considerable crop losses may be caused by parasitic fungi. These
diseases may be controlled by many different chemicals. Several preparations containing copper have been used for various blights and mildews
which are caused by fungi. When applied in high concentrations for many
years, as in apple orchards, the soil may become heavily contaminated with
copper so that the soil fauna is affected, and worms (for instance) are
eliminated, but little damage seems to be done to the mature trees and no
serious losses of the copper to contaminate surrounding areas have been
reported. Various organic fungicides (e.g. Captan) are very effective, particularly in horticulture, but as their total usage is not great as compared
with substances like MCPA and as they are generally rather expensive
(and therefore used with some care) no serious cases of pollution have
apparently occurred.

Various compounds using mercury are used as fungicides. Calomel
(HgCl), which itself is comparatively non-toxic so that it was formerly a
popular purgative prescribed for human consumption, has been used in
horticulture in fairly large amounts. Organomercury compounds are now
used as cereal seed dressings to such an extent that, in Britain, seed corn
is normally so dressed unless the customer stipulates otherwise. These
seed dressings are used prophylactically, i.e. they are used to prevent
common fungal infections, whether or not they are likely to cause damage
in a specific field.

Mercury can be an important environmental pollutant. It is stored and
accumulated by animals and may be transmitted and concentrated in food
chains. There have been many cases where fish and molluscs have accumulated quite high levels of mercury, and even deaths have occurred in men
eating the fish and molluscs. However, the serious cases of contamination
seem to have arisen from the much more massive industrial usage (e.g.
wood pulp processing), than from the comparatively small amounts used
in agriculture (see §6.2, p. 42).

Seed dressings are unlikely to contaminate the soil seriously. Only some 100 g of active ingredient are applied to a hectare, and the total addition of mercury, per square metre of soil, is in the region of a milligram. This is often negligible in comparison with the natural level of mercury in the soil.

Some contamination of wild birds occurs when dressed seed is eaten. In Britain the main dressings have been in the form of phenyl mercury compounds, which are not very toxic to vertebrates, and feeding experiments with pheasants suggest that they would never be likely to take up a dangerous dose. In some European countries methyl mercury seed dressings have been more widely used; these are much more toxic, and have been shown to kill birds accidentally. Even in Britain, contamination is not negligible, particularly as concentration in predators eating the seed-eating birds does occur, and may approach the level where physiological effects can be detected.

Different mercury compounds differ greatly in toxicity. One particular danger of this metal is that even where one of the least dangerous compounds, say calomel, is liberated into the environment, this may be transformed by bacterial action to the very toxic methyl form. This could occur within the gut of a ruminant, where methylization is known to occur, or in the bottom of a lake or in a swamp. For this reason every effort should be made to reduce the levels of mercury. However, this applies more to industry than to agriculture, and it should be remembered that the vast bulk of mercury, in soils or in the sea, occurs as a result of the natural weathering of mercury-containing rocks. However, though it occurs naturally, such mercury is as dangerous as that accidentally distributed by human actions.

7.3 Insecticides

Many poisons have been used to kill insect pests. Pests of fruit trees have been gassed with cyanide and sprayed with arsenic or with dinitro-orthocresol (DNOC). Nicotine was widely used, particularly in glasshouses, where it is as poisonous to insects as to cigarette smokers. Naturally occurring substances like the fish poison Derris have been found to be as effective against human lice and the Raspberry beetle. Any of these substances could be a pollutant and have harmful side effects, but in practice serious environmental pollution by insecticides is a post (1945) war problem, and dates from the widespread use of new chemicals which have been synthesized by man. With few exceptions, these synthetic insecticides have been much less acutely poisonous to man and other vertebrates than some of the the substances which they replaced, but the great persistence of some chemicals has produced new problems.

The insecticides in wide use today fall into two main groups, the organophosphorus and the organochlorine compounds. Some of the organophosphorus insecticides are the exceptions to the rule that synthetic insecticides are relatively non-poisonous to man. Parathion (diethyl p-nitrophenyl phosphorothionate) and TEPP (tetraethylpyrophosphate) are very poisonous, they may only be applied by operators using protective clothing, and they have caused thousands of accidental deaths and numerous others when taken deliberately as suicide drugs. However, though they have caused environmental damage when carelessly applied, they are unstable and soon break down to relatively harmless products and so they cannot cause long-lasting pollution.

Today, particularly in Britain, the very toxic organophosphorus insecticides are little used, as they have been largely replaced by safer substances. These include Malathion (o,o-dimethyl S-1,2-di(ethoxycarbonyl) ethyl phosphorodithioate) which is less than a thousandth as toxic as parathion, and various systemic insecticides such as Schradan (bis(tetramethyl-phosphorodiamidic) anhydride) and Demeton Methyl or 'Metasystox' (dimethyl S-2-ethylthioethyl phosphorothiolate) which are translocated through the plant and so are valuable in controlling aphids and other sap-feeding pests without directly endangering beneficial insects which do not feed on the plants. Carbamates, chemically distinct, but similar in action and persistence to organophosphates, are increasingly being used. As a whole they also do not cause lasting pollution. Today chemists are looking for substances which have a longer period of activity, and some of these may be more acutely toxic than, for instance, Malathion, but we are now aware of environmental dangers and bear them in mind when assessing new organophosphorus or carbamate insecticides.

There is, however, one particular danger from organophosphorus insecticides. They have what is, in effect, a cumulative action. Malathion, for instance, appears non-toxic to man and other mammals because they have a mechanism which de-toxifies it. It has been found that a previous, apparently harmless exposure to another organophosphorus compound, e.g. Parathion, may have affected this de-toxifying system, and then Malathion may do unexpected damage. This means that considerable care must be taken if even apparently safe amounts of these chemicals are to be used repeatedly.

7.4   Organochlorine Compounds

It is the stable and persistent organochlorine insecticides which have caused, and which are still causing, environmental contamination which at least sometimes is of sufficient magnitude to deserve the description of pollution. The insecticides most concerned are DDT (1,1-bis(p-chloro-phenyl)-2,2,2-trichloroethane), BHC or Lindane (1,2,3,4,5,6-hexachloro-cyclohexane) and members of the Cyclodiene Group, including Dieldrin or HEOD (1,2,3,4,10,10-hexachloro-6,7-epoxy-1,4,4a,5,6,7,8,8a-octa-hydro-exo-1,4-endo-5,8-dimethanonaphthalene), Aldrin or HHDN (1,2,3,4,10,10-hexachloro-1,4,4a,5,8,8a-hexahydro-exo-1,4-endo-5,8-

dimethanonaphthalene), and Heptachlor (1,4,5,6,7,10,10-heptachloro-4,7,8,9-tetrahydro-4,7-methyleneindene).

The insecticidal properties of DDT were recognized in 1939, and during the 1939-45 war malarial mosquitoes and typhus-bearing lice were controlled by it. After the war production was greatly increased, and DDT was widely used in controlling both medical and agricultural pests. It was found to be remarkably non-toxic to man, so that there is no case of a human death from the proper use of this insecticide. There is no doubt that DDT has saved millions of lives from malaria and other insectborne diseases in tropical countries, and that it has saved many from starvation by increasing food production by destroying crop pests. It was only when it began to be overused, and when Dieldrin, which is rather more toxic to vertebrates, was found to be killing many forms of wildlife, that concern about these chemicals began to be felt.

There is little doubt that DDT can now be detected almost everywhere in the world. It occurs in air and in rainwater, at levels of a few parts in 10^{11} and at rather higher amounts in the fat of birds and fish even in such unexpected places as the Antarctic. These low levels are not in themselves apparently having any biological effects, so they cannot be considered as pollution in the ordinary sense, though they are a cause for concern as we do not understand how they arose, or what is the route by which the insecticides got into the environment. The main worry, however, is lest concentration should occur, raising the levels to those which are indeed harmful. This occurs because organochlorine pesticides are very sparingly soluble in water, but extremely soluble in fat, in which they may often be stored. The stored insecticide may do little harm, but when the fat is metabolized and the chemical is released into the bloodstream, its effects may be dramatic.

There have undoubtedly been many cases of harmful pollution by organochlorines, with the death of non-target organisms. Thus in the United States massive amounts equal to over 30 kg per hectare of DDT were used against the beetle carrying the spores of the Dutch Elm Disease; this almost eliminated the American Robin mainly because it ate worms which had accumulated the insecticide. In Britain there is little direct evidence of unintentional deaths to birds or mammals from acute DDT poisoning, but when dieldrin was used as a seed-dressing to protect young plants from attack from the Wheat Bulb Fly larva, this seed was eaten by pigeons, pheasants and other birds, which died in large numbers. When the corpses were eaten by predatory birds and by foxes, these suffered fatal secondary poisoning. Some of the predators died from eating only one or two heavily contaminated prey, others concentrated the substances of many sublethal doses over a period from many apparently normal prey.

It is easy to recognize the effects of doses which are so high as to be fatal. The difficulty comes in assessing the possible damage from the more general and lower levels of contamination. Thus our food in Britain contains small but measurable amounts of organochlorine insecticides, though not high enough to be considered harmful (see Table 2). Nevertheless we do concentrate these substances, so that the levels in human fat are in the region of two parts per million. Sometimes people get the wrong idea that this process of concentration continues indefinitely. This is not the case. In Britain the levels of contamination of food have remained approximately the same for the last six or so years and so have the (higher) levels in human body fat. There is a 'plateau' effect in concentration, and there is actually some evidence that British levels are now beginning to fall as levels in the food decrease also. In the U.S.A., where food is more heavily contaminated, the human body burden is four or five times as high as in Britain. In Israel and India, where DDT is even more widely used, even higher body levels of 25 ppm or above have been found.

Table 2   Levels of some insecticides in food in Britain in 1968 (parts per million)

	BHC	Dieldrin	DDT and its metabolites
Beef fat	0.04	0.03	0.03
Mutton kidney fat	0.17	0.21	0.16
Butter	0.08	0.03	0.06
Milk	0.002	0.001	0.002
Eggs (when DDT is used in houses)	0.03	< 0.01	up to 1.70
Flour	0.018	0.001	0.009

There is no evidence that these levels in man—up to 25 ppm in their fat, much lower in the tissues—are having any physiological effects. Volunteers who over many months have deliberately ingested large amounts of DDT, and factory workers who have done so intentionally over periods of years, have had fat levels of 100 ppm or more, again without showing any clinical symptoms such as have been observed where DDT has been mistaken for flour and consumed on that scale.

Levels of organochlorine residues in wildlife are often higher than in man. It is sometimes difficult to evaluate these residues, as many different chemicals are involved. This is partly because chemicals are transformed within animal tissues. Thus DDT is largely turned into DDE, in a process of detoxification. In fact DDE is still insecticidal and poisonous to verte-brates, though less than DDT, except that certain processes in the breeding animal may be more rather than less severely affected.

There is good circumstantial evidence that the sublethal amounts of DDT (and DDE) as well as those of dieldrin have had their effects on wildlife. Thus it was shown that, in the late 1950s, there was a dramatic reduction in the thickness of eggshells of those predatory birds which accumulated the greatest body burden. Peregrines became rare, and

almost ceased to breed in southern Britain, in the areas where intensive
cereal growing was practised (Figs. 7–1 and 7–2). In Scotland Golden
Eagles became sufficiently contaminated to affect breeding, though the
population of adults (which may live for 30 years or more) has apparently
not been significantly reduced. It is also encouraging to be able to report
that, since the use of dieldrin as a seed dressing and a sheep dip has been
curtailed, a significant recovery in the breeding success of both hawks and

Fig. 7–1   Peregrines in Britain in 1961. In each region the first figure shows
the percentage of territories occupied, the second (in box) the percentage in
which young were reared. (After D. A. Ratcliffe, from MELLANBY, 1967)

eagles has occurred, parallel with a fall in the level of dieldrin in their tissues. The effects of sublethal levels of organochlorines have been confirmed by laboratory experiments which show that they affect the thyroid and the calcium metabolism, with the result that abnormal egg-shell development takes place. It seems that, in Britain, we have been fortunate enough to avoid the serious environmental damage which has occurred, with heavier pesticide usage, in the United States and other

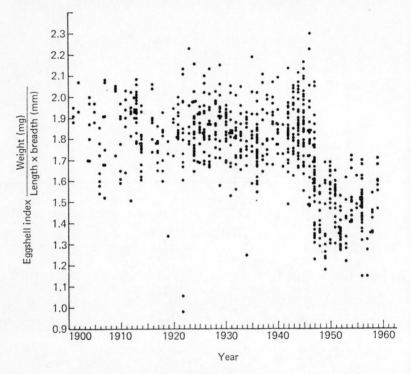

Fig. 7–2   Change in time of eggshell index (relative weight) of the pere-
grine in Britain. (From Ratcliffe, D. A. (1970). *J. appl. Ecol.*
7, 67–115)

countries, but that the margin of safety has not been great. Thus there is good reason to stop using these persistent chemicals as soon as possible, even if they may so far have done less damage than is often imagined.

I have pointed out that the danger of organochlorine insecticides is that they may be reconcentrated in living systems. One often hears it suggested that this commonly occurs step by step in food chains, thus if the chemical is applied to the soil, it may be taken up by worms, passed on to worm-eating birds, and finally concentrated (perhaps to dangerous levels) in hawks preying on them. This process is probably less important than is often imagined. It does occur but seldom builds up to anything approaching lethal levels in terrestrial food chains. The real danger is from the few individuals with really high rates of contamination.

In water, however, the situation is different. Fish and shrimps are able to concentrate pesticides either in solution or when absorbed onto minute particles in suspension, by a factor of 10^3 or even of 10^4. Thus they do indeed build up, in their tissues, levels of organochlorines which are harmful and even lethal. This was seen in the Rhine in 1968, when another organochlorine, Endosulfan (6,7,8,9,10,10-hexachloro-1,5,5a,6,9,9a-hexa-hydro-6,9-methano-2,4,3-benzo e dioxathiepin 3-oxide) was accidentally spilled into the river. Within a few days millions of fish died, though the initial level may have been as low as one part in ten million, a concentration which would have been harmless to man had he drunk the water.

In estuaries and waters near the shore, insecticide levels may similarly be high enough to be concentrated by fish and marine invertebrates to significant levels. This view is supported by the finding that many marine birds have substantial levels, and that seals and porpoises from the Baltic have had as much as 55 ppm in their blubber. Though not lethal, this is likely to have had some deleterious effect.

There is no doubt that there is some DDT in the body of our oceans, though the level must be below one part in 10^{12}, as it has not been measured in analyses with this degree of accuracy. Concern has been expressed lest this insecticide should harm the phytoplankton, and upset the oxygen balance of our atmosphere. Experiments have indeed shown that DDT will inhibit photosynthesis of plankton, but only at concentrations some 10^3 or more times those at present existing. As world usage of DDT (and of other organochlorine insecticides) is falling and not rising, really severe oceanic pollution of this kind seems unlikely.

My general conclusion is that organochlorine pesticides have caused severe local incidence of pollution; that some widespread contamination does exist, and that, particularly in water, all danger has not been eliminated. However, I believe that if the present trend to discontinue their use continues, then future risks are not great. However, we must continue to be on the look out for the side effects of the chemicals which replace them.

Appendix

A.1 Practical exercises on pollution

I have received many enquiries about possible exercises on pollution, its measurement and the assessment of its effects, from teachers and lecturers who wish to include such work in their curriculum, as well as from members of the public. Unfortunately work of this kind is seldom simple. Ecological studies, which are the most rewarding, usually take a long time, and require a detailed knowledge of the animals and plants concerned. Thus students may have heard that lichens are particularly susceptible to the effects of sulphur dioxide. When they visit an area where pollution is high, they may be surprised to find a rich growth of such a lichen as the species *Lecanora conizaoides*, which is able to flourish where the more susceptible species are entirely absent. This observation may lead them to learn to identify at least the commoner lichens, and this will allow them to make significant observations, but at the outset few of even the more ecologically minded teachers have this particular expertise.

Ideally, I would wish to be able to suggest practicable means by which the students could themselves make meaningful quantitative measurements of specific pollutants, and then follow up these measurements with biological studies which can be correlated with the chemical and physical observations. This is only possible in a limited number of cases, but sometimes sources of information about specific pollutants already exist, and much can be learned by visiting laboratories or treatment works. However, considerable preparation should precede such visits. It is hoped that the reading of this booklet may contribute to such preparation. Scientists are usually very happy to try to help the *informed* enquirers. They do not react with enthusiasm to the student who writes: 'I have to do a project on pollution. Please send me all the literature'.

This appendix, then, is mainly devoted to suggestions for lines of work which may be profitable. But those interested must be prepared to do a good deal of thinking and of preparatory study if their results are to have any significance.

A.2 Air pollution

I have already indicated that the study of air pollution is not simple, partly because the levels of pollutants, even in the worst areas, fluctuate so widely. 'Spot' measurements at a particular time, even if these were easily practicable, would have limited interest, as the observer might miss occasional, and lethal, emissions at much higher levels. Most observations give measurements related to the total emission over a period, often 24 hours, and this may be more significant, though it does not make it possible to quantify the effects of short bursts of high level pollution in distinction to longer productions of lower levels which will give the same mean figures.

The Warren Spring Laboratory at Stevenage in Hertfordshire is responsible for a Co-operative Investigation measuring the levels of smoke and
sulphur dioxide in some 1500 sites throughout Great Britain. This network
gives far more detailed information than is available for any other country.
Most of the stations are operated by local authorities or such organizations
as the Central Electricity Generating Board. You should obtain the latest
copy of *The Investigation of Air Pollution* from H.M. Stationery Office, or
from your local library, which should be able to borrow it for you.
Although the Warren Spring Laboratory itself is most helpful to enquirers, do not approach them directly for this sort of generalized information. If your local authority operates a centre, find out who is responsible
and contact them, and perhaps a useful visit to the site, together with a
lecture on the subject from whoever locally is most concerned, will form
a good introduction to the subject.

Should you wish to set up your own station, there are various possi-
bilities. The Open University may be able to help. They have devised a
kit which has already been successfully used by some thousands of their
students. Communicate with the Marketing Division, Open University,
Walton, Bletchley, Bucks. Some schools have themselves constructed
apparatus similar to that used in the Warren Spring network, after examin-
ing the machines in general use. This constructional work, using such
items as scrap gas meters, often arouses the interest of mechanically gifted
students. A kit 'Clean Air Research Pack' has been produced by the
Advisory Centre for Education and may be obtained from them (32
Trumpington Street, Cambridge) for 97p (post free). It enables some
chemical measurements to be made and also elementary determinations
of lichen distribution (see below).

Some quite unsophisticated measurements can be of value. In very smoky
areas (fortunately now increasingly uncommon) the deposition onto pieces
of white cloth may be compared at different sites. Rain may be collected
and analyzed—the pH in some places, where SO_2 is high, will be affected.
Such simple measurements often arouse interest among students, and
stimulate those most concerned to take their studies further.

The biological effects of air pollution can be studied in three ways.
First by surveys of the distribution of sensitive plants which are in fact the
assessment of the results of 'experiments' which have already been made.
Secondly, organisms can be introduced into areas where pollution is
suspected, and the effects observed. Thirdly, organisms can be exposed
(in glasshouses or other confined spaces) to controlled levels of pollutants.

(1) Lichens are known to be particularly susceptible to SO_2. Mosses
and liverworts may also be affected. Lichen surveys in and around towns
have proved rewarding, but, as already mentioned, require a good deal of
expertise. Before making such surveys you are advised to read some of the
original scientific papers on the subject. These include:

GILBERT, O. L. (1968). Bryophytes as indicators of air pollution in the Tyne
Valley. *New Phytol.* **67**, 15.

GILBERT, O. L. (1970). Further studies on the effect of sulphur dioxide on lichens and bryophytes. *New Phytol.* **69**, 605–627.

GILBERT, O. L. (1970). A biological scale for the estimation of sulphur dioxide pollution. *New Phytol.* **69**, 529–634.

HAWKSWORTH, D. L. and ROSE, F. (1970). Qualitative scale for estimating sulphur dioxide air pollution in England and Wales using epiphytic lichens. *Nature, Lond.* **227**, 145–148.

The best way to start learning to identify lichens is to find someone who can do so, and let him show you the technique. You will also require the recent book by U. K. Duncan, *Introduction to British Lichens* (1970), T. Buncle and Co. Ltd., Arbroath, 292p.

The following useful ideas for surveys have been kindly suggested by Dr. Gilbert:

1. Map the distribution of common wall top species such as *Parmelia saxatilis* (or even *Parmelia* sp.), *Grimmia pulvinata, Lecanora muralis*, down a suspected gradient of increasing pollution. Then map a common epiphytic lichen, i.e. *Parmelia sulcata* (or *Parmelia* sp.), *Evernia prunastri, Pertusaria* sp., or bryophyte, i.e. *Orthotrichum* sp., down the same gradient. Interpret the results. Does pH have an effect on survival? Does species of tree affect survival?

2. Map the distribution of *Xanthoria parietina* down a suspected gradient of increasing pollution on

   i) Calcareous substrata, i.e. asbestos roofs, concrete, limestone
   ii) Eutrophicated sandstone walls
   iii) Base of trees
   iv) Trunk of trees 1 m above the ground.

Interpret the results.

3. Collect the psocid *Mesopsocus unipunctatus* from standard trees, i.e. mature free standing ash trees in the open, down a gradient of increasing pollution. Is there any relationship between the number of individuals per unit area of trunk and the cover of epiphytes? Is there any sign of melanism? (see paper by GILBERT, O. L. (1971). Some indirect effects of air pollution on bark living invertebrates. *J. appl. Ecol.* **8**, 77–84).

4. Simply mapping the distribution of any common sensitive lichen or bryophyte (assemblages would be too hard) onto a 1 inch O.S. map using mapping pins will bring out the distribution of air pollution round a town or factory complex.

There are difficulties in all this. One has to be careful to standardize the habitat examined and of course in an industrial belt there may be no lichens worth mapping, within easy reach of the school. Perhaps the students could scan old floras to see what has disappeared and thus realize that pollution and habitat destruction often both act together to impoverish the flora. Town schools could investigate the distribution of *Lecanora conizaoides* and *Pleurococcus* together with the distribution of psocids on trees.

5. If new sources of pollution are opening up nearby, one can make a

tracing of local wall top lichen communities onto clear polythene sheeting and watch them disappear over the years.

(2) Different species and varieties of many plants differ in their susceptibility to air pollutants. Unfortunately the best 'pictorial atlas' is expensive, and is only obtainable from America. It is

Recognition of Air Pollution Injury to Vegetation: a Pictorial Atlas. Edited by JACOBSON, JAY S. and CLYDE HILL, A. Informative Report No. 1, Air Pollution Control Association, Pittsburgh, Pennsylvania (1970). Price: $15.

Susceptible varieties of plants can be grown in areas where air pollution is suspected. It is advisable to grow the plants in standard soil in pots, for the soil in an area where there has been long-standing pollution may itself be contaminated and this will affect growth. Table 3 shows some of the varieties of cultivated plant known to be susceptible.

Table 3 Varieties of plant known to be susceptible to some types of aerial pollutants

Plant	Injured by
Gladiolus, particularly var. Snow Prince	SO_2
Garden pea	SO_2
Zinnia	SO_2
Petunia	Ozone
Gladiolus	Fluoride
Maize	Fluoride

(3) Some of the plants listed in Table 3 may be grown in a glasshouse or enclosed with polythene, or better still in a bench-top growth chamber, and then exposed to a current of air containing a known concentration of SO_2 or other pollutant. A crude experiment, able to demonstrate gross damage, may be made by burning sulphur. The calculation on p. 69 shows that 1.5 mg of sulphur should give a concentration of 1 ppm of SO_2 in a cubic metre of air. However such a level will not be maintained, and it is probably necessary to burn about a gram of sulphur daily in a small glasshouse to obtain, in a space of perhaps two weeks, noticeable lesions on plant leaves. Various experiments on these lines may be devised.

More scientifically a known volume of air containing an exact level of pollutant should be circulated into the glasshouse. If the polluted air is introduced at a slightly greater pressure than that of the outside atmosphere, reasonably constant conditions can be obtained. Experiments of this kind are obviously only possible in institutions with sophisticated facilities and where constant supervision is available.

A.3 Water pollution

Much of our inland water is obviously and grossly polluted, so work on this subject is less difficult than that on the air. Much can be learned from visits to local sewage works, where the staff are often prepared to demonstrate the methods used, including analytical techniques for studying effluents. Before such expeditions lecturers and teachers without first hand experience of sewage works should make careful preparation and, if possible a preliminary visit. Excursions may also be possible to the laboratories run by the various River Authorities, where excellent work on the study of pollutants may be in progress. Here again I must stress the importance of preparation by teachers and students before approaching such authorities, in order to get the most out of a visit.

Water analysis is not difficult. Details may be obtained from two of the Handbooks produced in connection with the International Biological Programme.

GOLTERMAN, H. L. (1969). *Methods of Chemical Analysis of Fresh Waters.* IBP Handbook No. 8. Blackwell Scientific Publications, Oxford, p. 172.

VOLLENWEIDER, RICHARD A. (1969). *A Manual on Methods of Measuring Primary Production in Aquatic Environments.* IBP Handbook No. 12, Blackwell Scientific Publications, Oxford, p. 213.

Biological studies may be of every type of sophistication. A 'Water Pollution Kit' which has already been used in a very interesting survey by some 10 000 children throughout Britain may be obtained, price 75p, from the Advisory Centre for Education, 32 Trumpington Street, Cambridge. This forms an admirable introduction to the subject, enabling even quite young children to learn something about the main pollutants, and their effects as demonstrated by a number of 'indicator animals'. Comparisons should be made between apparently clean and dirty waters, and the effects of effluents, e.g. from sewage plants or factories, can be studied. Many will wish to graduate from the necessarily simple observations made using the kit to more complex studies. The kit contains a list of books for further study.

Various experiments with water from streams, ponds and rivers can be made. The following are among those kindly suggested by Professor R. W. Edwards.

1. Some measure of the growth-promoting properties of waters for algae can be determined using culture of algae (e.g. *Sceletonema* and *Chlorella*) incubated under standard light conditions in filtered water samples taken from rivers and lakes (+ dilutions of such waters in a range of standard waters, e.g. distilled or de-ionized). There are many variants to this approach for it can be extended by adding specific nutrients. Growth can be measured by filtering off the algae after a given interval of time (say, one week) and determining chlorophyll concentrations using a colorimeter with appropriate filter—or even by weighing the algal crop.

2. Toxicity experiments taking samples of effluents or river waters using invertebrates. Cladocerans like *Daphnia* are very good as they only need small beakers and with planktonic animals one can record the time it takes before they fall to the bottom. These experiments can also be conducted with known concentrations of poisons. This is a very good lead into toxicity

studies LC_{50} (i.e. the concentration lethal to half the population), slope functions, and so on.

3. Toxicity experiments in rivers etc. using small nylon-covered cages anchored to the bed containing invertebrates like *Gammarus*.

A.4 Thermal pollution

If access to water being heated (e.g. by a power station) is possible, temperature measurements can be made, and can be related to studies of the flora and fauna, much as in other investigations of water pollution.

A.5 Radiation

This is not a subject suitable for investigation except by experts under controlled conditions.

A.6 Pollution of the sea

Studies may be made of bacteria near sewage outfalls. The beach flora and fauna of such areas can be compared with unpolluted regions.

A.7 Pesticides

Fortunately the long lasting organochlorine pesticides are not available for field tests, so long-term pollution studies with them cannot now be made. Analyses of animal tissues for pesticides are usually outside the capabilities of any but the expert.

Many ecological experiments on plants and animals may still be devised. Thus broad beans can be given different insecticidal treatments, and the number of aphids and of their natural enemies may be counted. Long term effects of herbicides on the floral composition of grass sward may be studied. Observations on the numbers of many species of wildlife on farms following different practices and different pesticide uses can be rewarding.

Postscript Delta Directory of Environmental Literature and Teaching Aids, compiled by Carol Johnson and Jacqui Smith, The Council for Environmental Education, 26 Bedford Square, London WC1B 3HU.

This contains information about study kits. It also deals with books, films, filmstrips, slides, posters, games and workcards on the environment. Many refer to pollution. The Directory is produced in a ready-punched, loose-leaf form, and additional sheets will be issued to keep it up to date.

Further Reading

This list gives publications dealing with general problems of pollution in greater depth than was possible in the text. The Appendix contains additional references relating to methods of measuring and of recording the effects of specific pollutants.

ALEXANDER, P. (1957). *Atomic Radiation and Life.* Penguin Books, London. 239 pp.

CARSON, R. (1963). *Silent Spring.* Hamish Hamilton, London. 304 pp.

COLEMAN-COOKE, J. (1965). *The Harvest that Kills.* Odhams, London, 208 pp.

COMMITTEE OF THE ROYAL COLLEGE OF PHYSICIANS OF LONDON ON SMOKING AND ATMOSPHERIC POLLUTION. (1970). *Summary and Report on Air Pollution and its Effect on Health.* Pitman Medical and Scientific, London. 80 pp.

EUROPEAN CONGRESS ON THE INFLUENCE OF AIR POLLUTION ON PLANTS AND ANIMALS, 1st. *Air Pollution.* Centre for Agricultural Publishing and Documentation, Wageningen. 415 pp.

GILL, C., BOOKER, F. and SOPER, T. (1967). *The Wreck of the Torrey Canyon.* David and Charles, Newton Abbot. 128 pp.

HARTLEY, G. S. and WEST, T. F. (1969). *Chemicals for Pest Control.* Pergamon Press, London. 316 pp.

GOODMAN, G. T., EDWARDS, R. W. and LAMBERT, J. M. (1965). *Ecology and the Industrial Society.* A Symposium of the British Ecological Society, Swansea, 13–16 April, 1964. Blackwell, Oxford. 395 pp.

HYNES, H. B. N. (1960). *The Biology of Polluted Waters.* Liverpool University Press. 202 pp.

JONES, F. G. W. and JONES, M. (1964). *Pests of Field Crops.* Edward Arnold, London. 406 pp.

JONES, J. R. ERICHSEN. (1964). *Fish and River Pollution.* Butterworth, London. 203 pp.

LAVERTON, S. (1962). *The Profitable Use of Farm Chemicals.* Oxford University Press, London. 96 pp.

MACAN, T. T. and WORTHINGTON, E. B. (1951). *Life in Lakes and Rivers.* New Naturalist No. 15, Collins, London. 272 pp.

MELLANBY, K. (1967). *Pesticides and Pollution.* New Naturalist No. 50, Collins, London. 212 pp.

PIRIE, A., Ed. (1958). *Fallout.* Revised edition. MacGibbon and Kee, London. 176 pp.

ROSE, J., Ed. (1969). *Technological Injury.* Gordon and Breach, London. 224 pp.

ROYAL COMMISSION ON ENVIRONMENTAL POLLUTION, First Report, February 1971. H.M.S.O., Cmnd. 4585, London. 52 pp.

RUDD, R. L. (1965). *Pesticides and the Living Landscape.* Faber and Faber, London. 320 pp.

THE ENVIRONMENTAL POLLUTION PANEL. (1965). *Restoring the Quality of our Environment.* Report of the President's Science Advisory Committee, November, 1965, 317 pp.

YAPP, W. B. (1959). *The Effects of Pollution on Living Material.* Symposia of the Institute of Biology, No. 8, London. 154 pp.